The
Tuning
of the
World

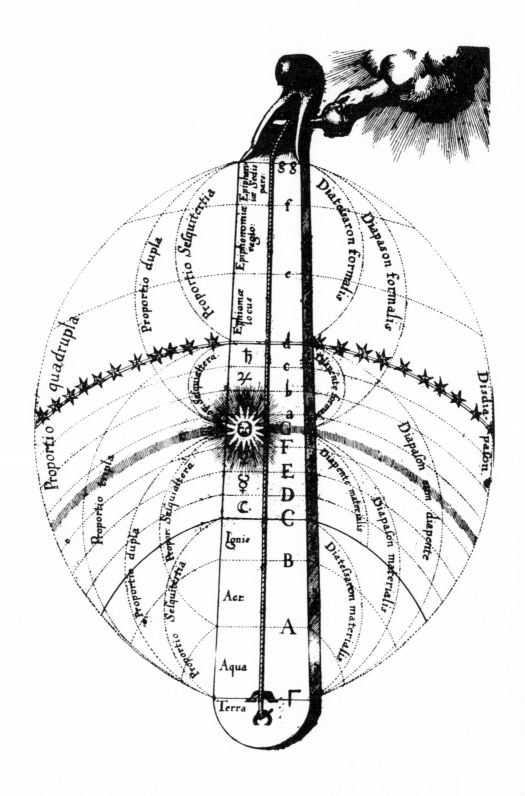

R. MURRAY SCHAFER

The Tuning of the World

Alfred A. Knopf   New York   1977

THIS IS A BORZOI BOOK
PUBLISHED BY ALFRED A. KNOPF, INC.

Copyright © 1977 by R. Murray Schafer
All rights reserved under International and Pan-American Copyright Conventions.
Published in the United States by Alfred A. Knopf, Inc., New York.
Distributed by Random House, Inc., New York.

Library of Congress Cataloging in Publication Data

Schafer, R. Murray.   The tuning of the world.

Includes bibliographical references and index.
1. Music—Acoustics and physics.   2. Sound.
3. Music—Philosophy and aesthetics.   I. Title.
ML3805.s3   781'.1   76-49508
ISBN 0-394-40966-3

Manufactured in the United States of America
First Edition

Grateful acknowledgment is made to the following for permission to reprint previously published material:

E. J. Brill: Chart from "The Spectographic Analysis of Sound Signals of the Domestic Fowl" by N. Collias and M. Joos, originally published in *Behaviour*, V.

Cambridge University Press: Chart from *Bird-Song* by W. H. Thorpe, copyright © 1961, p. 63. Also, six lines from *Theocritus*, edited and translated by A. S. F. Gow, copyright 1950, Vol. I, *Idyll I*.

Clarke, Irwin & Company Limited: Excerpts from *Hundreds and Thousands* by Emily Carr, copyright © 1966 by Clarke, Irwin & Company Limited.

Harcourt Brace Jovanovich, Inc.: Excerpts from *Technics and Civilization* by Lewis Mumford, copyright 1934.

Indiana University Press: Excerpts from *Formalized Music* by Iannis Xenakis, copyright © 1971, pp. 8–9.

New Directions Publishing Corporation and Faber and Faber Limited: 28 lines from *Canto II*

The decorative illustrations used in this book are drawn from several sources:

Page ii, woodcut of "The Tuning of the World" from Robert Fludd's *Utriusque Cosmi Historia* (1617).

Page 13, Etruscan wall painting of a pipe player with a bird from Triclinium, Tarquinia, Italy.

Page 69, nineteenth-century factory scene published in *Das Buch der Erfindungen, Gewerbe und Industrien,* vol. VI (Leipzig and Berlin, 1874).

Page 101, London street cries, an eighteenth-century engraving by William Hogarth.

Page 121, examples of graphic notation systems used by architects, geographers and acousticians.

Page 203, illustration of an imaginary sound magnification system from *Phonurgia Nova* by Athanasius Kircher (1673).

To my co-workers on
the World Soundscape Project

Contents

Preface

Ever since I began studying the acoustic environment, it has been my hope to gather my work together into one book which might serve as a guide for future research. This book accordingly borrows extensively from many of my previous publications, in particular the booklets *The New Soundscape* and *The Book of Noise* and the several documents of the World Soundscape Project, especially the essay *The Music of the Environment* and our first comprehensive field study, *The Vancouver Soundscape*. But it tries to build this fugitive material into a more careful arrangement.

As evidence has come in from more distant sources and as I have reflected further or been provoked by my fellow researchers, many early assumptions have been revised or abandoned. The present book is as definitive as it can be at the present time, but since only God knows for sure, it must still be regarded as tentative.

Much of the material of this book was revealed through an international study entitled the World Soundscape Project, which many agencies helped to fund. To my immediate co-workers on the Project I owe a great debt of gratitude for countless stimulating meetings and discussions. It is as much their book as it is mine, for they read it, criticized it and provided both facts and encouragement. In particular I wish to thank Hildegard Westerkamp, Howard Broomfield, Bruce Davis, Peter Huse and Barry Truax. Jean Reed, now my wife, was a special help in checking sources, reading numerous drafts and tolerating the many moods of the author.

Numerous scholars in different disciplines have encouraged soundscape studies. Many have read portions of this book and have provided helpful commentaries. Others have suggested new angles of investigation or have sent material from abroad which could not otherwise have been obtained. In particular I wish to thank the following scholars: Professor

Kurt Blaukopf and Dr. Desmond Mark of the Institute for Music, Dance and Theatre, Vienna; G. S. Métraux and Anny Malroux of UNESCO, Paris; Dr. Philip Dickinson, Department of Bioengineering, University of Utah; Professor John Large, Institute of Sound and Vibration Research, University of Southampton; Dr. David Lowenthal, Department of Geography, University College, London; Dr. Peter Ostwald, Langley Porter Neuropsychiatric Institute, University of California; Marshall McLuhan, Centre for Culture and Technology, University of Toronto; Michel P. Philippot, l'Institut National de l'Audiovisuel, Paris; Dr. Catherine Ellis, University of Adelaide; Professor John Paynter, University of York; Professor Jean-Jacques Nattiez, l'Université de Montréal; and Professor Pat Shand, University of Toronto.

I am especially indebted to Yehudi Menuhin for his constant encouragement of soundscape research and to Dr. Otto Laske for his valuable commentaries on my text.

The World Soundscape Project could lay no claim to its title without numerous reports and verifications from many countries. For supplying special information, or for helping to translate it, I thank: David Ahern, Carlos Araujo, Renata Braun, Junko Carothers, Mieko Ikegame, Roger Lenzi, Beverley Matsu, Judith Maxie, Albert Mayr, Marc Métraux, Walter Otoya, John Rimmer, Thorkell Sigurbjörnsson, Turgut Var and Yngve Wirkander. Nick Reed deserves special thanks for valuable library research.

For typing numerous drafts of the manuscript I am thankful to Pat Tait, Janet Knudson and Linda Clark. When an author keeps changing his mind, typists have the hardest job of all.

R. MURRAY SCHAFER
Vancouver, August, 1976

*The
Tuning
of the
World*

Introduction

Now I will do nothing but listen . . .
I hear all sounds running together, combined,
    fused or following,
Sounds of the city and sounds out of the city, sounds
    of the day and night. . . .
                    WALT WHITMAN, *Song of Myself*

The soundscape of the world is changing. Modern man is beginning to inhabit a world with an acoustic environment radically different from any he has hitherto known. These new sounds, which differ in quality and intensity from those of the past, have alerted many researchers to the dangers of an indiscriminate and imperialistic spread of more and larger sounds into every corner of man's life. Noise pollution is now a world problem. It would seem that the world soundscape has reached an apex of vulgarity in our time, and many experts have predicted universal deafness as the ultimate consequence unless the problem can be brought quickly under control.

In various parts of the world important research is being undertaken in many independent areas of sonic studies: acoustics, psychoacoustics, otology, international noise abatement practices and procedures, communications and sound recording engineering (electroacoustics and electronic music), aural pattern perception and the structural analysis of language and music. These researches are related; each deals with aspects of the world soundscape. In one way or another researchers engaged on these various themes are asking the same question: what is the relationship between man and the sounds of his environment and what happens when those sounds change? Soundscape studies attempt to unify these various researches.

Noise pollution results when man does not listen carefully. Noises are the sounds we have learned to ignore. Noise pollution today is being resisted by noise abatement. This is a negative approach. We must seek a way to make environmental acoustics a *positive* study program. Which sounds do we want to preserve, encourage, multiply? When we know this, the boring or destructive sounds will be conspicuous enough and we will know why we must eliminate them. Only a total appreciation of the acoustic environment can give us the resources for improving the orchestration of the world soundscape. For many years I have been fighting for ear cleaning in schools to eliminate audiometry in factories. Clairaudience not ear muffs. It is an idea over which I do not wish to exercise permanent ownership.*

The home territory of soundscape studies will be the middle ground between science, society and the arts. From acoustics and psychoacoustics we will learn about the physical properties of sound and the way sound is interpreted by the human brain. From society we will learn how man behaves with sounds and how sounds affect and change his behavior. From the arts, particularly music, we will learn how man creates ideal soundscapes for that other life, the life of the imagination and psychic reflection. From these studies we will begin to lay the foundations of a new interdiscipline—acoustic design.

From Industrial Design to Acoustic Design The most important revolution is aesthetic education in the twentieth century was that accomplished by the Bauhaus, that celebrated German school of the twenties. Under the leadership of architect Walter Gropius, the Bauhaus collected some of the great painters and architects of the time (Klee, Kandinsky, Moholy-Nagy, Mies van der Rohe), together with craftsmen of distinction. At first it seemed disappointing that the graduates of this school did not rise to rival their mentors as artists. But the purpose of the school was different. From the interdisciplinary synergy of faculty skills a whole new study field was created, for the school invented the subject of industrial design. The Bauhaus brought aesthetics to machinery and mass production.

It devolves on us now to invent a subject which we might call acoustic design, an interdiscipline in which musicians, acousticians, psychologists, sociologists and others would study the world soundscape together in order to make intelligent recommendations for its improvement. This study would consist of documenting important features, of noting differences, parallels and trends, of collecting sounds threatened with extinction, of studying the effects of new sounds before they are indiscriminately

*For definitions of *ear cleaning, clairaudience* and other special terms, see the glossary.

released into the environment, of studying the rich symbolism sounds have for man and of studying human behavior patterns in different sonic environments in order to use these insights in planning future environments for man. Cross-cultural evidence from around the world must be carefully assembled and interpreted. New methods of educating the public to the importance of environmental sound must be devised. The final question will be: is the soundscape of the world an indeterminate composition over which we have no control, or are *we* its composers and performers, responsible for giving it form and beauty?

Orchestration Is a Musician's Business Throughout this book I am going to treat the world as a macrocosmic musical composition. This is an unusual idea but I am going to nudge it forward relentlessly. The definition of music has undergone radical change in recent years. In one of the more contemporary definitions, John Cage has declared: "Music is sounds, sounds around us whether we're in or out of concert halls: cf. Thoreau." The reference is to Thoreau's *Walden,* where the author experiences in the sounds and sights of nature an inexhaustible entertainment.

To define music merely as *sounds* would have been unthinkable a few years ago, though today it is the more exclusive definitions that are proving unacceptable. Little by little throughout the twentieth century, all the conventional definitions of music have been exploded by the abundant activities of musicians themselves. First with the huge expansion of percussion instruments in our orchestras, many of which produce nonpitched and arhythmic sounds; then through the introduction of aleatoric procedures in which all attempts to organize the sounds of a composition rationally are surrendered to the "higher" laws of entropy; then through the opening-out of the time-and-space containers we call compositions and concert halls to allow the introduction of a whole new world of sounds outside them (in Cage's *4'33" Silence* we hear only the sounds external to the composition itself, which is merely one protracted caesura); then in the practices of *musique concrète,* which inserts any sound from the environment into a composition via tape; and finally in electronic music, which has revealed a whole gamut of new musical sounds, many of them related to industrial and electric technology in the world at large.

Today all sounds belong to a continuous field of possibilities lying *within the comprehensive dominion of music.* Behold the new orchestra: the sonic universe!

And the musicians: anyone and anything that sounds!

Dionysian Versus Apollonian Concepts of Music It is easier to see the responsibilities of the acoustical engineer or the audiologist toward the world soundscape than to understand the precise manner in which the contemporary musician is supposed to attach himself to this vast theme, so I am going to grind my axe on this point for a moment longer.

There are two basic ideas of what music is or ought to be. They may be seen most clearly in two Greek myths dealing with the origin of music. Pindar's twelfth Pythian Ode tells how the art of aulos playing was invented by Athena when, after the beheading of Medusa, she was touched by the heart-rending cries of Medusa's sisters and created a special *nomos* in their honor. In a Homeric hymn to Hermes an alternative origin is mentioned. The lyre is said to have been invented by Hermes when he surmised that the shell of the turtle, if used as a body of resonance, could produce sound.

In the first of these myths music arises as subjective emotion; in the second it arises with the discovery of sonic properties in the materials of the universe. These are the cornerstones on which all subsequent theories of music are founded. Characteristically the lyre is the instrument of Homer, of the epos, of serene contemplation of the universe; while the aulos (the reed oboe) is the instrument of exaltation and tragedy, the instrument of the dithyramb and of drama. The lyre is the instrument of Apollo, the aulos that of the Dionysian festivals. In the Dionysian myth, music is conceived as internal sound breaking forth from the human breast; in the Apollonian it is external sound, God-sent to remind us of the harmony of the universe. In the Apollonian view music is exact, serene, mathematical, associated with transcendental visions of Utopia and the Harmony of the Spheres. It is also the *anāhata* of Indian theorists. It is the basis of Pythagoras's speculations and those of the medieval theoreticians (where music was taught as a subject of the quadrivium, along with arithmetic, geometry and astronomy), as well as of Schoenberg's twelve-note method of composition. Its methods of exposition are number theories. It seeks to harmonize the world through acoustic design. In the Dionysian view music is irrational and subjective. It employs expressive devices: tempo fluctuations, dynamic shadings, tonal colorings. It is the music of the operatic stage, of *bel canto,* and its reedy voice can also be heard in Bach's Passions. Above all, it is the musical expression of the romantic artist, prevailing throughout the nineteenth century and on into the expressionism of the twentieth century. It also directs the training of the musician today.

Because the production of sounds is so much a subjective matter with modern man, the contemporary soundscape is notable for its dynamic hedonism. The research I am about to describe represents a reaffirmation of music as a search for the harmonizing influence of sounds in the world about us. In Robert Fludd's *Utruisque Cosmi Historia* there is an illustration entitled "The Tuning of the World" in which the earth forms the body of an instrument across which strings are stretched and are tuned by a divine hand. We must try once again to find the secret of that tuning.

Music, the Soundscape and Social Welfare In Hermann
Hesse's *The Glass Bead Game* there is an arresting idea. Hesse claims to be
repeating a theory of the relationship between music and the state from
an ancient Chinese source: "Therefore the music of a well-ordered age is
calm and cheerful, and so is its government. The music of a restive age is
excited and fierce, and its government is perverted. The music of a decay-
ing state is sentimental and sad, and its government is imperiled."

Such a theory would suggest that the egalitarian and enlightened reign
of Maria Theresa (for instance, as expressed in her unified criminal code
of 1768) and the grace and balance of Mozart's music are not accidental.
Or that the sentimental vagaries of Richard Strauss are perfectly consistent
with the waning of the same Austro-Hungarian Empire. In Gustav Mahler
we find, etched in an acid Jewish hand, marches and German dances of
such sarcasm as to give us a presentiment of the political *dance macabre*
soon to follow.

The thesis is also borne out well in tribal societies where, under the
strict control of the flourishing community, music is tightly structured,
while in detribalized areas the individual sings appallingly sentimental
songs. Any ethnomusicologist will confirm this. There can be little doubt
then that music is an indicator of the age, revealing, for those who know
how to read its symptomatic messages, a means of fixing social and even
political events.

For some time I have also believed that the general acoustic environment of a society can be read as an indicator of social conditions which
produce it and may tell us much about the trending and evolution of that
society. Throughout this book I will suggest many such relationships, and
though it is probably in my nature to do this emphatically, I hope the
reader may continue to regard the method as valid even if some of the
equations seem disagreeable. They are all open to further testing.

The Notation of Soundscapes (Sonography) The sound-
scape is any acoustic field of study. We may speak of a musical composition as a soundscape, or a radio program as a soundscape or an acoustic
environment as a soundscape. We can isolate an acoustic environment as
a field of study just as we can study the characteristics of a given landscape.
However, it is less easy to formulate an exact impression of a soundscape
than of a landscape. There is nothing in sonography corresponding to the
instantaneous impression which photography can create. With a camera it
is possible to catch the salient features of a visual panorama to create an
impression that is immediately evident. The microphone does not operate
this way. It samples details. It gives the close-up but nothing corresponding to aerial photography.

Similarly, while everyone has had some experience reading maps, and many can draw at least significant information from other schematics of the visual landscape, such as architects' drawings or geographers' contour maps, few can read the sophisticated charts used by phoneticians, acousticians or musicians. To give a totally convincing image of a soundscape would involve extraordinary skill and patience: thousands of recordings would have to be made; tens of thousands of measurements would have to be taken; and a new means of description would have to be devised.

A soundscape consists of events *heard* not objects *seen*. Beyond aural perception is the notation and photography of sound, which, being silent, presents certain problems that will be discussed in a special chapter in the Analysis section of the book. Through the misfortune of having to present data on silent pages, we will be forced to use some types of visual projection as well as musical notation, in advance of this discussion, and these will only be useful if they assist in opening ears and stimulating clairaudience.

We are also disadvantaged in the pursuit of a historical perspective. While we may have numerous photographs taken at different times, and before them drawings and maps to show us how a scene changed over the ages, we must make inferences as to the changes of the soundscape. We may know exactly how many new buildings went up in a given area in a decade or how the population has risen, but we do not know by how many decibels the ambient noise level may have risen for a comparable period of time. More than this, sounds may alter or disappear with scarcely a comment even from the most sensitive of historians. Thus, while we may utilize the techniques of modern recording and analysis to study contemporary soundscapes, for the foundation of historical perspectives, we will have to turn to earwitness accounts from literature and mythology, as well as to anthropological and historical records.

Earwitness   The first part of the book will be particularly indebted to such accounts. I have always attempted to go directly to sources. Thus, a writer is trustworthy only when writing about sounds directly experienced and intimately known. Writing about other places and times usually results in counterfeit descriptions. To take an obvious instance, when Jonathan Swift describes Niagara Falls as making "a terrible squash" we know he never visited the place; but when Chateaubriand tells us that in 1791 he heard the roar of Niagara eight to ten miles away, he provides us with useful information about the ambient sound level, against which that of today could be measured. When a writer writes uncounterfeitingly about directly apprehended experiences, the ears may sometimes play tricks on the brain, as Erich Maria Remarque discovered in the trenches during the First World War when he heard shells exploding about him followed by the rumble of the distant guns that fired them. This aural illusion is perfectly accountable, for as the shells were traveling at super-

sonic speeds they arrived in advance of the sounds of their original detona-
tions; but only someone trained in acoustics could have predicted this. *All
Quiet on the Western Front* is convincing because the author was there. And
we trust him when he describes other unusual sound events—for instance,
the sounds made by dead bodies. "The days are hot and the dead lie
unburied. We cannot fetch them all in, if we did we should not know what
to do with them. The shells will bury them. Many have their bellies
swollen up like balloons. They hiss, belch, and make movements. The
gases in them make noises." William Faulkner also knew the noise of
corpses, which he described as "little trickling bursts of secret and murmur-
ous bubbling."

In such ways is the authenticity of the earwitness established. It is a
special talent of novelists like Tolstoy, Thomas Hardy and Thomas Mann
to have captured the soundscapes of their own places and times, and such
descriptions constitute the best guide available in the reconstruction of
soundscapes past.

Features of the Soundscape What the soundscape analyst must
do first is to discover the significant features of the soundscape, those
sounds which are important either because of their individuality, their
numerousness or their domination. Ultimately some system or systems of
generic classification will have to be devised, and this will be a subject for
the third part of the book. For the first two parts it will be enough to
categorize the main themes of a soundscape by distinguishing between
what we call *keynote sounds, signals* and *soundmarks.* To these we might add
archetypal sounds, those mysterious ancient sounds, often possessing
felicitous symbolism, which we have inherited from remote antiquity or
prehistory.

Keynote is a musical term; it is the note that identifies the key or
tonality of a particular composition. It is the anchor or fundamental tone
and although the material may modulate around it, often obscuring its
importance, it is in reference to this point that everything else takes on its
special meaning. Keynote sounds do not have to be listened to consciously;
they are overheard but cannot be overlooked, for keynote sounds become
listening habits in spite of themselves.

The psychologist of visual perception speaks of "figure" and
"ground," the figure being that which is looked at while the ground exists
only to give the figure its outline and mass. But the figure cannot exist
without its ground; subtract it and the figure becomes shapeless, nonexis-
tent. Even though keynote sounds may not always be heard consciously,
the fact that they are ubiquitously there suggests the possibility of a deep
and pervasive influence on our behavior and moods. The keynote sounds
of a given place are important because they help to outline the character
of men living among them.

The keynote sounds of a landscape are those created by its geography and climate: water, wind, forests, plains, birds, insects and animals. Many of these sounds may possess archetypal significance; that is, they may have imprinted themselves so deeply on the people hearing them that life without them would be sensed as a distinct impoverishment. They may even affect the behavior or life style of a society, though for a discussion of this we will wait until the reader is more acquainted with the matter.

Signals are foreground sounds and they are listened to consciously. In terms of the psychologist, they are figure rather than ground. Any sound can be listened to consciously, and so any sound can become a figure or signal, but for the purposes of our community-oriented study we will confine ourselves to mentioning some of those signals which *must* be listened to because they constitute acoustic warning devices: bells, whistles, horns and sirens. Sound signals may often be organized into quite elaborate codes permitting messages of considerable complexity to be transmitted to those who can interpret them. Such, for instance, is the case with the *cor de chasse,* or train and ship whistles, as we shall discover.

The term *soundmark* is derived from landmark and refers to a community sound which is unique or possesses qualities which make it specially regarded or noticed by the people in that community. Once a soundmark has been identified, it deserves to be protected, for soundmarks make the acoustic life of the community unique. This is a subject to be taken up in Part Four of the book, where the principles of acoustic design will be discussed.

I will try to explain all other soundscape terminology as it is introduced. At the end of the book there is a short glossary of terms which are either neologistic or have been used idiosyncratically, in case doubt exists at any point in the text. I have tried not to use too many complex acoustical terms, though a knowledge of the fundamentals of acoustics and a familiarity with both musical theory and history is presupposed.

Ears and Clairaudience   We will not argue for the priority of the ear. In the West the ear gave way to the eye as the most important gatherer of information about the time of the Renaissance, with the development of the printing press and perspective painting. One of the most evident testaments of this change is the way in which we have come to imagine God. It was not until the Renaissance that God became portraiture. Previously he had been conceived as sound or vibration. In the Zoroastrian religion, the priest Srosh (representing the genius of hearing) stands between man and the pantheon of the gods, listening for the divine messages, which he transmits to humanity. *Samā* is the Sufi word for audition or listening. The followers of Jalal-ud-din Rumi worked themselves into a mystical trance by chanting and whirling in slow gyrations. Their dance is thought by some scholars to have represented the solar system, recalling also the deep-rooted mystical belief in an extraterrestrial music, a Music of the Spheres, which the attuned soul may at times hear. But these exceptional powers of hearing, what I have called clairaudience, were not attained effortlessly. The poet Saadi says in one of his lyric poems:

> I will not say, my brothers, what *samā* is
> Before I know who the listener is.

Before the days of writing, in the days of prophets and epics, the sense of hearing was more vital than the sense of sight. The word of God, the history of the tribe and all other important information was heard, not seen. In parts of the world, the aural sense still tends to predominate.

> . . . rural Africans live largely in a world of sound—a world loaded with direct personal significance for the hearer—whereas the western European lives much more in a visual world which is on the whole indifferent to him. . . . Sounds lose much of this significance in western Europe, where man often develops, and must develop, a remarkable ability to disregard them. Whereas for Europeans, in general, "seeing is believing," for rural Africans reality seems to reside far more in what is heard and what is said. . . . Indeed, one is constrained to believe that the eye is regarded by many Africans less as a receiving organ than as an instrument of the will, the ear being the main receiving organ.

Marshall McLuhan has suggested that since the advent of electric culture we may be moving back to such a state again, and I think he is right. The very emergence of noise pollution as a topic of public concern testifies to the fact that modern man is at last becoming concerned to clean the sludge out of his ears and regain the talent for clairaudience—clean hearing.

A Special Sense   Touch is the most personal of the senses. Hearing and touch meet where the lower frequencies of audible sound pass over to tactile vibrations (at about 20 hertz). Hearing is a way of touching at a distance and the intimacy of the first sense is fused with sociability whenever people gather together to hear something special. Reading that sentence an ethnomusicologist noted: "All the ethnic groups I know well have in common their physical closeness and an incredible sense of rhythm. These two features seem to co-exist."

The sense of hearing cannot be closed off at will. There are no earlids. When we go to sleep, our perception of sound is the last door to close and it is also the first to open when we awaken. These facts have prompted McLuhan to write: "Terror is the normal state of any oral society for in it everything affects everything all the time."

The ear's only protection is an elaborate psychological mechanism for filtering out undesirable sound in order to concentrate on what is desirable. The eye points outward; the ear draws inward. It soaks up information. Wagner said: "To the eye appeals the outer man, the inner to the ear." The ear is also an erotic orifice. Listening to beautiful sounds, for instance the sounds of music, is like the tongue of a lover in your ear. Of its own nature then, the ear demands that insouciant and distracting sounds would be stopped in order that it may concentrate on those which truly matter.

Ultimately, this book is about sounds that matter. In order to reveal them it may be necessary to rage against those which don't. In Parts One and Two I will take the reader on a long excursion of soundscapes through history, with a heavy concentration on those of the Western world, though I will try to incorporate material from other parts of the world whenever it has been obtainable. In Part Three the soundscape will be subjected to critical analysis in preparation for Part Four, where the principles of acoustic design will be outlined—at least as far as they can be determined at the moment.

All research into sound must conclude with silence—a thought which must await its development in the final chapters. But the reader will clearly sense that this idea also links the first part of the book to the last, thus uniting an undertaking that is above all lyrical in character.

One final warning. Although I will at times be treating aural perception and acoustics as if they were abstractable disciplines, I do not wish to forget that the ear is but one sense receptor among many. The time has come to move out of the laboratory into the field of the living environment. Soundscape studies do this. But even they must be integrated into that wider study of the total environment in this not yet best of all possible worlds.

PART ONE

~~~~~~

First Soundscapes

In those days men's ears heard sounds
whose angelic purity cannot be conjured
up again by any amount of science or magic.

HERMAN HESSE, *The Glass Bead Game*

ONE

~~~~~~

The Natural Soundscape

Voices of the Sea   What was the first sound heard? It was the caress of the waters. Proust called the sea "the plaintive ancestress of the earth pursuing, as in the days when no living creature existed, its lunatic immemorial agitation." The Greek myths tell how man arose from the sea: "Some say that all gods and all living creatures originated in the stream of Oceanus which girdles the world, and that Tethys was the mother of all his children."

The ocean of our ancestors is reproduced in the watery womb of our mother and is chemically related to it. Ocean and Mother. In the dark liquid of ocean the relentless masses of water pushed past the first sonar ear. As the ear of the fetus turns in its amniotic fluid, it too is tuned to the lap and gurgle of water. At first it is the submarine resonance of the sea, not yet the splash of wave. But then

> ... the waters little by little began to move, and at the movement of the waters the great fish and the scaly creatures were disturbed, and the waves began to roll in double breakers, and the beings that dwell in the waters were seized with fear and as the breakers rushed together in pairs the roar of the ocean grew loud, and the spray was lashed into fury, and garlands of foam arose, and the great ocean opened to its depths, and the waters rushed hither and thither, the furious crests of their waves meeting this way and that.

Waves whipped into surf, pelting the first rocks as the amphibian ascends from the sea. And although he may occasionally turn his back on the waves, he will never escape their atavistic charm. "The wise man delights in water," says Lao-tzu. The roads of man all lead to water. It is the

fundamental of the original soundscape and the sound which above all
others gives us the most delight in its myriad transformations.

At Oostende the strand is wide, with a scarcely perceptible rake across
to the hotels, so that standing there one has the impression that the sea
in the distance is higher than the beach and that sooner or later everything
will be lifted away to oblivion by an enormous soft tidal wave. Totally
otherwise is the Adriatic at Trieste, where the mountains leap into the
ocean with an angular energy and the angry fists of the waves bounce
noisily off rocks like India rubber balls. At Oostende the nexus of land is
gentle in both vista and tone.

There are no rocks on which to sit at Oostende and so one walks along
for miles, south with the waves in the right ear, and north with the waves
in the left ear, filling an atavistic consciousness with the full-frequencied
throb of water. All roads lead to water. Given the chance, probably all men
would live at the edge of the element, within earshot of its moods night
and day. We wander from it but the departure is temporary.

Day after day one walks along the strand, listening to the indolent
splashing of the wavelets, gauging the gradual crescendo to the heavier
treading and on to the organized warfare of the breakers. The mind must
be slowed to catch the million transformations of the water, on sand, on
shale, against driftwood, against the seawall. Each drop tinkles at a different pitch; each wave sets a different filtering on an inexhaustible supply
of white noise. Some sounds are discrete, others continuous. In the sea the
two fuse in primordial unity. The rhythms of the sea are many: infrabio-
logical—for the water changes pitch and timbre faster than the ear's resolving power to catch its changes; biological—the waves rhyme with the
patterns of heart and lung and the tides with night and day; and suprabiological—the eternal inextinguishable presence of water. "Observe measures," says Hesiod in *Works and Days;* "I will show you the measures of
the much-thundering sea."

para thina polyphloisboio thalassēs

says Homer (*Iliad*, I:34), catching onomatopoeically the splendid armies of
waves on the sea beach and their recession. *Canto II* of Ezra Pound begins,

> And poor old Homer blind, blind, as a bat,
> Ear, ear for the sea-surge . . .

The love of ocean has profound sources and they are recorded in a vast
maritime literature of East and West. When water watches the history of
the tribe, fingers of ocean grasp the epic. The prime material over which
the *Odyssey* is strung is the ocean. The agrarian Hesiod, living in Boeotia,
"far away from the sea and its tossing waters," cannot avoid the lure of
the ocean.

> For fifty days, after the turn
> of the summer solstice,

> when the wearisome season of the hot weather
>   goes to its conclusion
> then is the timely season for men to voyage.

The Norsemen knew the ferocity of the ocean. When they sailed, "waves roared against the sides of the ship, it sounded just as if boulders were being clashed together." The alliterative verse of the Eddas is poetry for oarsmen. The repeated consonants of each half-line pin the accents of the verse to each stroke and return of the oar.

> Splashing oars raced   iron rattled
> shield rang on shield   as the Vikings rowed,
> cutting the waves   at the King's command,
> farther and farther   the fleet sped on.

> When the crested waves   of Kolga's sister
> crashed on the keels   the sound that came
> was the boom of surf   that breaks on rocks.

Across the world, in tropical northern Australia, the waves were more gentle.

> Waves coming up: high waves coming up against the rocks,
> Breaking, shi! shi!
> When the moon is high with light upon the waters:
> Spring tide; tide flowing to the grass,
> Breaking, shi! shi!
> In its rough waters, the young girls bathe.
> Hear the sound they make with their hands as they play!

Any visitor to the seashore will find the recital of the waves remarkable, but only the maritime poet, with the ostinato of the sea in his ear from birth to grave, can measure precisely the systole and diastole of waves and tides. Ezra Pound spent much of his life moving from one coast of the Italian peninsula to the other—from Rapallo to Venice. His *Cantos* open on the sea, play out much of their dialectic at its edge, move away and then return. Where Scott Fitzgerald, a visitor to the Mediterranean, had heard merely "the small exhausted wa-*waa* of the waves," Pound gives us the fluctuations of the water with instinctive authority.

> Lithe turning of water,
>   sinews of Poseidon,
> Black azure and hyaline,
>   glass wave over Tyro,
> Close cover, unstillness,
>   bright welter of wave-cords,
> Then quiet water,
>   quiet in the buff sands,
> Sea-fowl stretching wing-joints,
>   splashing in rock-hollows and sand-hollows

In the wave-runs by the half-dune;
Glass-glint of wave in the tide-rips against sunlight,
          pallor of Hesperus,
Grey peak of the wave,
          wave, colour of grape's pulp,
Olive grey in the near,
          far, smoke grey of the rock-slide,
Salmon-pink wings of the fish-hawk
          cast grey shadows in water,
The tower like a one-eyed great goose
          cranes up out of the olive-grove,
And we have heard the fauns chiding Proteus
          in the smell of hay under the olive-trees,
And the frogs singing against the fauns
          in the half-light.
And . . .

The sea is the keynote sound of all maritime civilizations. It is also a fertile sonic archetype. All roads lead back to water. We shall return to the sea.

The Transformations of Water

Water never dies. It lives forever reincarnated as rain, as bubbling brooks, as waterfalls and fountains, as swirling rivers and deep sulking rivers.

A mountain stream is a chord of many notes strung out stereophonically across the path of the attentive listener. The continuous sound of water from Swiss mountain streams can be heard miles across a silent valley. When a stream leaps down a hundred-meter cascade in the Rocky Mountains, there is tense quietness, almost like fear, followed by noisy excitement when it strikes the rocks below. The water of the English moors has none of this virtuosity; its arrangements are more subtle.

> The wanderer in this direction who should stand still for a few moments on a quiet night, might hear singular symphonies from these waters, as from a lampless orchestra, all playing in their sundry tones from near and far parts of the moor. At a hole in a rotten weir they executed a recitative; where a tributary brook fell over a stone breastwork they trilled cheerily; under an arch they performed a metallic cymballing; and at Durnover Hole they hissed.

The rivers of the world speak their own languages. The gentle murmur of the Merrimack River, "whirling and sucking, and lapsing downward, kissing the shore as it went," was a sleeping pill for Thoreau. For James Fenimore Cooper, the rivers of upstate New York often moved sluggishly into rocky caverns "producing a hollow sound, that resembled the concussions of a distant gun."

How different are the furious cataracts of the Nile at Atbara and Berber.

For the noise of battle cannot but arise when the river, among a thousand islands and rocks, forges its way onward in mile-long rapids. A Roman writer declared that the inhabitants emigrated because they lost their hearing, but the mighty voices of the Berbers prove to us today that necessity strengthens any organ, for their call carries over the rushing river from bank to bank, while white men can hardly hear each other at ten paces' distance.*

By contrast, on the still rivers of Siam, Somerset Maugham found a "sensation of exquisite peace," only occasionally broken by "the soft splash of a paddle as someone silently passed on his way home. When I awoke in the night I felt a faint motion as the houseboat rocked a little and heard a little gurgle of water, like the ghost of an Eastern music travelling not through space but through time." In Thomas Mann's *Death in Venice* the wasted and mournful waters of the canals form a tragic leitmotiv: "Water slapped gurgling against wood and stone. The gondolier's cry, half warning, half salute, was answered with singular accord from far within the silence of the labyrinth."

Water never dies and the wise man rejoices in it. No two raindrops sound alike, as the attentive ear will detect. Is then the sound of Persian rain like that of the Azores? In Fiji a summer rainstorm whips past in an enormous swirl taking less than sixty seconds, while in London it drones on as boring as a businessman's story. In parts of Australia it does not rain for two or more years. When it does, young children are sometimes frightened by the sound. On the Pacific coast of North America it rains gently but continuously on an average of 148 days each year. The Canadian painter Emily Carr describes it well:

The rain drops hit the roof with smacking little clicks, uneven and stabbing. Through the open windows the sound of the rain on the leaves is not like that. It is more like a continuous sigh, a breath always spending with no fresh intake. The roof rain rattles over our room's hollowness, strikes and is finished.

The tranquil timpani of West Coast rain is ambitionless, quite unlike the violent thunderstorms of the plains of Russia and central North America. In South Africa the rain is torrential: ". . . the thunder boomed out over-

*The Roman writer referred to is Pliny (*Natural History*, V. x. 54), who merely states that the cataracts were very noisy but does not claim that they caused deafness. A legend, nevertheless, seems to have grown up to this effect, for we find it mentioned in Bernardino Ramazzini's *Diseases of Workers* (*De Morbis Artificium*) of 1713, a work which is remarkable for being the first known study to mention industrial deafness.

head, and they could hear the rain rushing across the fields. In a moment it was drumming on the iron roof, with a deafening noise."

Geography and climate provide vernacular keynotes to the soundscape. In the vast northern areas of the earth the sound of winter is that of frozen water—of ice and snow. During the winter 30 to 50 percent of the surface of the earth is covered by snow for some length of time, and 20 to 30 percent of the land surface is snow covered for more than six months annually. Ice and snow form the keynotes of the northern hinterland as surely as the sea is the keynote of maritime life.

Ice and snow are tuned by the temperature. Virginia Woolf at Blackfriars heard the snow "slither and flop to the ground." But in Scandinavia, when the giant Hymir of *The Elder Edda* returned from hunting:

> Icicles clattered,
>   falling off his frozen beard.

In his poem *Orfano,* Giovanni Pascoli describes the slow flaking snow of Italy:

> *Lenta la neva fiocca, fiocca, fiocca.*

The sound of snow in barely freezing Italy is very different from that at 30 degrees below zero in Manitoba or Siberia. As one moves to the interiors of the great northern continents the soft padded step begins to crunch, then to squeak—even painfully. Boris Pasternak in *Doctor Zhivago* tells how felt boots in the Russian winter make "the snow screech angrily at each step."

While seascapes have enriched the languages of maritime peoples, cold-climate civilizations have invented different expressions, of which the numerous Eskimo words for snow is the most celebrated though by no means the only instance. *The Illustrated Glossary of Snow and Ice* contains 154 terms for snow and ice in English and matches them with terms in Danish, Finnish, German, Icelandic, Norwegian, Russian, French-Canadian and Argentinian Spanish. Many of the expressions—for instance, *permafrost, icebound, pack ice*—are absent from the vocabularies of other languages.

Snow absorbs sound and northern literature is full of descriptions of the silence of winter.

> In wintertime, the stillness, the absence of life or sound, is weird and oppressive. When the snow is on the ground, you may perceive indeed the footprints of animals, of birds, of deer, or occasionally of a bear, but you hear no sound, not a cry, not a whisper, not a rustle of a leaf. Sit down upon a fallen tree, and the silence becomes oppressive, almost painful. It is a relief even to hear at last the sough of the fall of the snow from the boughs of the cypress, the pine, or the yew, which stretch like dark horse-plumes high overhead.

When the snow is fresh and soft, even the traditional creaking of the runners of a sleigh are mute. ". . . we glided along over virgin snow which had come soft-footedly over night, in a motion, so smooth and silent as to suggest that wingless flight . . ." Even the cities were quiet.

> Nor is anything quite like the silence of a northern city at dawn on a winter morning. Occasionally there was a hiss of whisper and a brushing against the windows and I knew it was snow, but generally there was nothing but a throbbing stillness until the street cars began running up Côte des Neiges and I heard them as though they were winds blowing through old drains.

The destruction of the quiet northern winter by the jamming of snow-plows and snowmobiles is one of the greatest transmogrifications of the twentieth-century soundscape, for such instruments are destroying the "idea of North" that has shaped the temperament of all northern peoples and has germinated a substantial mythology for the world. The idea of North, at once austere, spacious and lonely, could easily throw fear into the heart (had not Dante refrigerated the center of his Hell?) but it could evoke intense awe, for it was pure, temptationless and silent. The technocrats of progress do not realize that by cracking into the North with their machinery, they are chopping up the integrity of their own minds, blacking the awe-inspiring mysteries with gas stations and reducing their legends to plastic dolls. As silence is chased from the world, powerful myths depart. That is to say, it becomes more difficult to appreciate the Eddas and sagas, and much that is at the center of Russian, Scandinavian and Eskimo literature and art.

The traditional winter of the North is remarkable for its stillness, but the spring is violent. At first there is a determined grinding of ice, then suddenly a whole river will rip down the center with a cannon shot and spring water will hurtle the ice downstream. When asked what he most loved about Russia, Stravinsky said, "The violent Russian spring that seemed to begin in an hour and was like the whole earth cracking."

Voices of the Wind   Among the ancients, the wind, like the sea, was deified. In *Theogony,* Hesiod tells how Typhoeus, the god of the winds, fought with Zeus, lost, and was banished to Tartaros, in the bowels of the earth. Typhoeus was a devious god. He possessed a hundred snake heads,

> and inside each one of these horrible heads
> there were voices
> that threw out every sort of horrible sound,
> for sometimes
> it was speech such as the gods
> could understand, but at other
> times, the sound of a bellowing bull,

> proud-eyed and furious
> beyond holding, or again like a lion
> shameless in cruelty,
> or again it was like the barking of dogs,
> a wonder to listen to,
> or again he would whistle
> so the tall mountains re-echoed to it.

The story is remarkable because it touches on one of the most interesting aural illusions. The wind, like the sea, possesses an infinite number of vocal variations. Both are broad-band sounds and within the breadth of their frequencies other sounds seem to be heard. The deceptiveness of the wind is also the subject of a tempestuous description by Victor Hugo. You must read this aloud in the original to feel the pressure of the language.

> Le vaste trouble des solitudes a une gamme; crescendo redoutable: le grain, la rafale, la bourrasque, l'orage, la tourmente, la tempête, la trombe: les sept cordes de la lyre des vents, les sept notes de l'abîme. . . . Les vents courent, volent, s'abattent, finissent, recommencent, planent, sifflent, mugissent, rient; frénétiques, lascifs, effrénés, prenant leurs aises sur la vague irascible. Ces hurleurs ont une harmonie. Ils font tout le ciel sonore. Ils soufflent dans la nuée comme dans un cuivre, ils embouchent l'espace, et ils chantent dans l'infini, avec toutes les voix amalgamées des clairons, des buccins, des oliphants, des bugles et des trompettes, une sorte de fanfare prométhéenne. Qui les entend écoute Pan.

The wind is an element that grasps the ears forcefully. The sensation is tactile as well as aural. How curious and almost supernatural it is to hear the wind in the distance without feeling it, as one does on a calm day in the Swiss Alps, where the faint, soft whistling of the wind over a glacier miles away can be heard across the intervening stillness of the valleys.

On the dry Saskatchewan prairie the wind is keen and steady.

> The wind could be heard in a more persistent song now, and out along the road separating the town from the prairie it fluted gently along the wires that ran down the highway. . . . The night wind had two voices; one that keened along the pulsing wires, the prairie one that throated long and deep.

Treeless and open, the prairies are an enormous wind harp, vibrating incessantly with "the swarming hum of the telephone wires." In the more sheltered English countryside, the wind sets the leaves shimmering in diverse tonalities.

> To dwellers in a wood almost every species of tree has its voice as well as its feature. At the passing of the breeze the fir-trees sob and moan no less distinctly than they rock; the holly whistles as it battles with

itself; the ash hisses amid its quiverings; the beech rustles while its flat boughs rise and fall. And winter, which modifies the note of such trees as shed their leaves, does not destroy its individuality.

Sometimes I ask students to identify moving sounds in the soundscape. "The wind," say some. "Trees," say others. But without objects in its path, the wind betrays no apparent movement. It hovers in the ears, energetic but directionless. Of all objects, trees give the best cues, shaking their leaves now on one side, now on the other as the wind brushes them.

Each type of forest produces its own keynote. Evergreen forest, in its mature phase, produces darkly vaulted aisles, through which sound reverberates with unusual clarity—a circumstance which, according to Oswald Spengler, drove the northern Europeans to try to duplicate the reverberation in the construction of Gothic cathedrals. When the wind blows in the forests of British Columbia, there is nothing of the rattling and rustling familiar with deciduous forests; rather there is a low, breathy whistle. In a strong wind the evergreen forest seethes and roars, for the needles twist and turn in turbine motion. The lack of undergrowth or openings into clearings keeps the British Columbia forests unusually free of animal, bird and insect life, a circumstance which produced an awesome, even sinister impression on the first white settlers. Emily Carr again:

> The silence of our Western forests was so profound that our ears could scarcely comprehend it. If you spoke your voice came back to you as your face is thrown back to you in a mirror. It seemed as if the forest were so full of silence that there was no room for sounds. The birds who lived there were birds of prey—eagles, hawks, owls. Had a song bird loosed his throat the others would have pounced. Sober-coloured, silent little birds were the first to follow settlers into the West. Gulls there had always been; they began with the sea and had always cried over it. The vast sky spaces above, hungry for noise, steadily lapped up their cries. The forest was different—she brooded over silence and secrecy.

The uneasiness of the early settlers with the forest, and their desire for space and sunlight, soon produced another keynote sound: the noise of lumbering. At first it was the woodsman's axe that was heard just beyond the ever-widening clearing. Later it was the cross-saw, and today it is the snarl of the chainsaw that resounds throughout the diminishing forest communities of North America.

Once, much of the world was covered with forest. The great forest is foreign and appalling, quite inimical to intruding life. The few references to nature in the early epics, the sagas and Anglo-Saxon poetry testify to this fact; they are either brief or dwell on its horrors. Even as late as Carl Maria von Weber (1786–1826) the forest was a place of darkness and evil, and his opera *Der Freischütz* is a celebration of goodness over the forces of evil, whose home is the forest. The hunting horn, which Weber used so brilliantly in his score, became the acoustic symbol by which the gloom of the forest was transpierced.

When man was fearful of the dangers of an unexplored environment, the whole body was an ear. In the virgin forests of North America, where vision was restricted to a few feet, hearing was the most important sense. The Leatherstocking Tales of Fenimore Cooper are full of beautiful and terrifying surprises.

> ... for, though the quiet deep of solitude reigned in that vast and nearly boundless forest, nature was speaking with her thousand tongues, in the eloquent language of night in a wilderness. The air sighed through ten thousand trees, the water rippled, and, at places, even roared along the shores; and now and then was heard the creaking of a branch, or a trunk, as it rubbed against some object similar to itself, under the vibrations of a nicely balanced body. . . . When he desired his companions, however, to cease talking, in the manner just mentioned, his vigilant ear had caught the peculiar sound that is made by the parting of a dried branch of a tree, and which, if his senses did not deceive him, came from the western shore. All who are accustomed to that particular sound will understand how readily the ear receives it, and how easy it is to distinguish the tread which breaks the branch from every other noise of the forest. . . . "Can the accursed Iroquois have crossed the river, already, with their arms, without a boat?"

The Miraculous Land   "What is the sound of a tree falling in the woods with no one there to hear it?" asks a student who has studied philosophy. It would be unimaginative to reply that it sounds merely like a tree falling in the woods, or even that it makes no sound at all. As a matter of fact, when a tree crashes in a forest and knows that it is alone, it sounds like anything it wishes—a hurricane, a cuckoo, a wolf, the voice of Immanuel Kant or Charles Kingsley, the overture to *Don Giovanni* or a delicate air blown on a Maori nose-flute. Anything it wishes, from past or distant future. It is even free to produce those secret sounds which man will never hear because they belong to other worlds. . . .

The demystification of the elements, to which many modern sciences have contributed, has turned much poetry into prose. Before the birth of the earth sciences, man lived on an enchanted earth. From a third-century *Treatise on Rivers and Mountains,* perhaps by Plutarch, we learn of a stone in Lydia called argrophylax which looks like silver:

> It is rather difficult to recognize it because it is intimately intermixed with the little spangles of gold which are found in the sands of the river. It has one very strange property. The rich Lydians place it under the threshold of their treasurehouses, and thus protect their stores of gold. For whenever any robbers come near the place, the stone gives

forth a sound like a trumpet and the would-be thieves, believing themselves to be pursued, flee and fall over precipices and thus come to a violent death.

In earlier times, all natural events were explained as miracles. An earthquake or a storm was a drama between the gods. When Sigurd killed the dragon Fafner, "the earth tremors were so violent that all the land round about shook." When the Giants stole Donner's thundering hammer

> his hair stood upright,   his beard shook with wrath,
> wild for his weapon   the god ˙groped around.

There was bound to be a mighty storm. When Zeus led the Greek gods against the Titans

> ... the infinite great sea
>   moaned terribly
> and the earth crashed aloud,
>   and the wide sky resounded. ...
> Now Zeus no longer held in his strength,
>   but here his heart filled
> deep with fury, and now he showed
>   his violence entire
> and indiscriminately. Out of the sky
>   and off Olympos
> he moved flashing his fires incessantly,
>   and the thunderbolts,
> the crashing of them and the blaze
>   together came flying, one after
> another, from his ponderous hand,
>   and spinning whirls of inhuman
> flame, and with it the earth,
>   the giver of life, cried out
> aloud as she burned, and the vast forests
>   in the fire screamed. ...

Donner and Zeus are still comprehensible gods even today. Thunder and lightning are among the most feared forces in nature. The sound is of great intensity and extreme frequency range, well outside the human scale of soundmaking. The gulf between men and the gods is great and often it has seemed as if a mighty noise was necessary to bridge it. Such a noise was that of the eruption of Vesuvius in A.D. 79 when, according to Dion Cassius's account, "the frightened people thought the Gyants were making war against heaven, and fansied they see the shapes and images of Gyants in the smoke, and heard the sound of their trumpets." The event was one of the soundmarks of Roman history.

Then the Earth began to tremble and quake, and the Concussions were so great that the ground seem'd to rise and boyl up in some places, and in others the tops of the mountains sunk in or tumbled down. At the same time were great noises and sounds heard, some were subterraneous, like thunder within the Earth; others above ground, like groans or bellowings. The Sea roar'd, The heavens ratled with a fearful noise, and then came a sudden and mighty crack, as if the frame of Nature had broke, or all the mountains of the Earth had faln down at once. . . .

Unique Tones   Every natural soundscape has its own unique tones and often these are so original as to constitute soundmarks. The most striking geographical soundmark I have ever heard is in New Zealand. At Tikitere, Rotorua, great fields of boiling sulphur, spread over acres of ground, are accompanied by strange underground rumblings and gurglings. The place is a pustular sore on the skin of the earth with infernal sound effects boiling up through the vents.

The volcanoes of Iceland produce something of the same effect, but moving back from them one is surprised by the change of sound effects.

At the crater itself there are thunderous, explosive sounds and even near the crater you can feel the ground shaking. The fatal walls of lava (2–3 meters high) inch out killing everything in their path. They are almost silent, but not quite, for listening carefully you can hear delicate, brittle snaps in the crust—dry clicks, like the fracturing of glass, spread out over several miles. When it meets wet land the lava also hisses in a suffocating sort of way. Otherwise all is nearly silent.

Even where there is no life, there can be sound. The ice fields of the North, for instance, far from being silent, reverberate with spectacular sounds.

Within three or four miles of the glaciers you begin to hear the cracking of massive ice packs. It sounds like distant thunder and recurs every five or six minutes. As you get closer you can distinguish between the initial crack, like a huge pane of glass being cracked, followed by the rumble of falling ice, and then the whole is reverberated distantly in the mountains.

Rivers of glacier water form tunnels underneath the ice. The falling ice inside these tunnels, the running water and the movement of mud and rocks create a noise which is amplified many times by the hollow structure and hits the observer on the surface with great force.

Nor is it silent below the earth's surface as Heinrich Heine discovered when he visited the mines of the Harz Mountains in 1824.

I did not reach the deepest section . . . the point I reached seemed deep enough,—a constant rumbling and roaring, sinister groaning of machinery, bubbling of subterranean springs, water trickling down, ev-

When Krakatoa exploded on the night of August 26, 1883,
the sound was reported heard over the area shaded here.

erywhere thick exhalations, and the miner's lamp flickering ever more
feebly in the lonesome night.

Apocalyptic Sounds   Perhaps the universe was created silently.

We do not know. The dynamics of the wonder which introduced our
planet were without human ears to hear them. But the prophets exercised
their imagination over the event. "In the beginning was the Word," says
John; God's presence was first announced as a mighty vibration of cosmic
sound. The prophets had a vision of the end also making a mighty noise.
References are especially plentiful in Judaic and Muslim prophecies.

> Howl ye; for the day of the Lord is at hand. . . . I will shake the
> heavens, and the earth shall remove out of her place, in the wrath of
> the Lord of hosts, and in the day of his fierce anger.

By the din of the drums of resurrection they have pressed tight their
two ears in terror.

They put their fingers in their ears against the thunderclaps, fearful
of death.

In the imagination of the prophets the end of the world was to be signaled
by a mighty din, a din more ferocious than the loudest sound they could
imagine: more ferocious than any known storm, more outrageous than any
thunder.

The loudest noise heard on this earth within living memory was the
explosion of the caldera Krakatoa in Indonesia on August 26 and 27, 1883.
The actual sounds were heard as far away as the island of Rodriguez, a
distance of nearly 4,500 kilometers, where the chief of police reported:
"Several times during the night . . . reports were heard coming from the
eastward, like the distant roars of heavy guns. These reports continued at
intervals of between three and four hours, until 3 p.m. on the 27th. . . ."
On no other occasion have sounds been perceived at such great distances,
and the area over which the sounds were heard on August 27 totaled
slightly less than one-thirteenth of the entire surface of the globe.

It is as difficult for the human being to imagine an apocalyptic noise
as it is for him to imagine a definitive silence. Both experiences exist in
theory only for the living since they set limits to life itself, though they
may become unconscious goals toward which the aspirations of different
societies are drawn. Man has always tried to destroy his enemies with
terrible noises. We shall encounter deliberate attempts to reproduce the
apocalyptic noise throughout the history of warfare, from the clashing of
shields and the beating of drums in ancient times right up to the Hiroshima
and Nagasaki atom bombs of the Second World War. Since that time
worldwide destruction has been lessened perhaps, but sonic destruction
has not, and it is disconcerting to realize that the ferocious acoustical
environment produced by modern civilian life derives from the same eschatological urge.

TWO

~~~~~~

The Sounds of Life

Bird-Song   One of the most beautiful miracles in all literature and mythology occurs in the midst of the brutalities of the *Saga of the Volsungs* when Sigurd, after slaying the dragon Fafner and tasting his blood, suddenly understands the language of the birds—a moment which Wagner used to great advantage in his opera *Siegfried.*

The language and song of the birds has been a subject of much study, though still today it is highly debatable whether the birds "sing" or "converse," in the customary sense of those terms. Nevertheless, no sound in nature has attached itself so affectionately to the human imagination as bird vocalizations. In tests in many countries we have asked listeners to identify the most pleasant sounds of their environment; bird-song appears repeatedly at or near the top of the list. And the history of effective bird imitations in music extends from Clément Janequin (d. *c.* 1560) to Olivier Messiaen (b. 1908).

Like birds themselves, bird vocalizations are of all types. A few are penetratingly loud. The call of the rufous scrubbird (*Atrichornis rufescens*) of Australia "is so intense that it leaves a sensation in one's ears." Other birds can at times dominate a soundscape because of their numbers. The bell minor bird (*Manorina melanophrys*) heard around Melbourne, with its persistent bell-like ring always sounding at approximately the same pitch (E♮ –F♮ –F♯), gives rise to a soundscape as dense as that created by cicadas, but different in that it maintains a certain spatial perspective; for the bird sounds issue from recognizable points, unlike the stridulations of the cicadas, which create a continual presence, seemingly without foreground or background.

In most parts of the world, bird-song is rich and varied, without being imperialistically dominating. Thus, St. Francis of Assisi adopted birds as symbolic of gentleness in much the same manner as his Muslim contemporary Jalal-ud-din Rumi adopted the reed flute for his mystic sect as a symbol of humility and simplicity in opposition to the vulgarity and opulence of his time. The symbolic importance of bird-song for both music and the soundscape is a subject to be returned to later.

The vocalizations of birds have often been studied in musical terms. In the early days ornithologists constructed charming words in no man's language to describe their sounds.

Hawfinch	*Deak . . . waree-ree-ree Tchee . . . tchee . . . tur-wee-wee*
Greenfinch	*wah-wah-wah-wah-chow-chow-chow-chow-tu-we-we*
Crossbill	*jibb . . . chip-chip-chip-gee-gee-gee-gee*
Great Titmouse	*ze-too, ze-too, p'tsee-ée, tsoo-ée, tsoo-ée ching-see, ching-see, deeder-deeder-deeder, biple-be-wit-se-diddle*
Pied Flycatcher	*Tchéetle, tchéetle, tchéetle diddle-diddle-dée; tzit-tzit-tzit, trui, trui, trui*
Mistlethrush	*tre-wir-ri-o-ee; tre-wir-ri-o-ee-o; tre-we-o-wee-o-wee-o-wit*
Corncrake	*crex-crex, krek-krek, rerp-rerp*
Common Snipe	*tik-tik-tik-tuk-tik-tuk-tik-tuk-chip-it; chick-chuck; yuk-yuk*

Musical notation was also used, and still is, by Olivier Messiaen, who has turned transcription into a complex art form. But despite the ingenuity of such work, bird vocalizations, with few exceptions, cannot be notated in musical terms. Many of the sounds uttered are not single tones but complex noises, and the high-frequency range and rapid tempo of many songs preclude their being transcribed in a notational system designed for the lower frequency ranges and slower tempi of human music. A more precise method of notation is that of the sound spectrograph and ornithologists are now using this method.

The structure of bird-song is often elaborate, for many birds are virtuoso performers. Some are also mimics. The Australian lyrebird is a superb mimic and its song often includes not only imitations of the songs of up to fifteen other species of birds, but also the neighing of horses, the sounds of cross-cut saws, car horns and factory whistles! The songs of many birds contain repetitive motifs, and though the function of the repetitions is often obscure, these melodic leitmotivs, variations and ex-

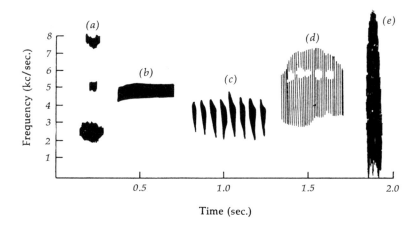

A sound spectrograph distinguishes clearly among bird notes having different tonal qualities: (a) nightingale note, very pure, with harmonics; (b) white-throated sparrow, clear whistle; (c) marsh warbler, musical trill; (d) clay-colored sparrow, toneless buzz; (e) budgerigar, noisy flight squawk.

pansions show certain similarities to melodic devices in music, such as those employed by the troubadours, or by Haydn and Wagner. In some details, the affective language of certain birds has been shown to bear a relationship to the shapes of human vocal and musical expression. For instance, the distress notes of chicks are composed of descending frequencies only, while ascending frequencies predominate in pleasure calls. The same general contours are present in man's expressions of sadness and pleasure.

But despite these similarities, it is obvious that to whatever extent the birds are deliberately communicating, it is for their own benefit rather than ours that their vocalizations are designed. Some men may puzzle over their codes, but most will be content merely to listen to the extravagant and astonishing symphony of their voices. Birds, like poems, should not mean, but be.

Bird Symphonies of the World
Each territory of the earth will have its own bird symphony, providing a vernacular keynote as characteristic as the language of the men who live there. In Paris, Victor Hugo listened to the birds in the Luxembourg Gardens during May, the month of mating.

> The quincunxes and flower-beds sent balm and dazzlement into the light, and the branches, wild in the brilliancy of midday, seemed

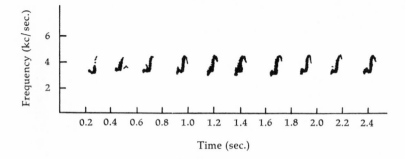

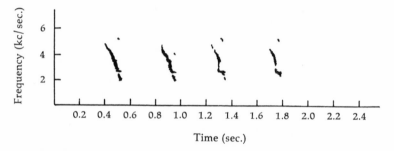

A sound spectrograph of the pleasure notes (above)
and distress call (below) of a three-day-old chick.

trying to embrace each other. There was in the sycamores a twittering of linnets, the sparrows were triumphal, and the woodpeckers crept along the chestnuts, gently tapping the holes in the bark. . . . This magnificence was free from stain, and the grand silence of happy nature filled the garden,—a heavenly silence, compatible with a thousand strains of music, the fondling tones from the nests, the buzzing of the swarms, and the palpitations of the wind.

Such rich polyphony is absent from the North American grasslands. On a plain near Pittsburgh a century ago, a German writer found "absolutely nothing. . . . Far and wide there was not a bird, nor a butterfly, nor the cry of an animal, not the hum of an insect." In the grasslands, sounds evaporated as if they had never been uttered. In the Russian Steppes, bird-song was also often isolated: "Everything might be dead; only above in the heavenly depths a lark is trilling and from the airy heights the silvery notes drop down upon adoring earth, and from time to time the cry of a gull or the ringing note of a quail sounds in the steppe." Occasionally only a single species is heard: "How enchanting this place was! Orioles kept making their clear three-note calls, stopping each time just long enough to let the countryside suck in the moist fluting sounds down to the last vibration." And in winter the birds blended with sleigh bells: "What could be more pleasant than to sit alone at the edge of a snowy field and listen to the chirping of the birds in the crystal silence of a winter's day, while somewhere far away in the distance sounded the bells of a passing troika —that melancholy lark of the Russian winter."

But in the jungles of Burma, such clarity was impossible to find, as Somerset Maugham discovered when he journeyed there. "The noise of the crickets and the frogs and the cries of the birds" produced a tremendous din, "so that till you become accustomed to it you may find it hard to sleep." "There is no silence in the East," Maugham concluded.

Ornithologists have not yet measured the statistical density of birds' singing in different parts of the world in sufficient detail for us to make objective comparisons—comparisons that would be helpful in mapping the complex rhythms of the natural soundscape. But ornithologists have done a lot of work on another subject of interest to soundscape researchers by classifying the types and functions of bird-song. Basically these are distinguished as follows:

> pleasure calls
> distress calls
> territorial-defense calls
> alarm calls
> flight calls
> flock calls
> nest calls
> feeding calls

Equivalents for many of these can be found in human soundmaking. To take some obvious examples: the territorial calls of birds are reproduced in automobile horn blowing, their alarm calls are reproduced in police sirens and their pleasure calls in the beach-side radio. In the territorial calls of birds we encounter the genesis of the idea of acoustic space, with which we will be much concerned later. The definition of space by acoustic means is much more ancient than the establishment of property lines and fences; and as private property becomes increasingly threatened in the modern world, it may be that principles regulating the complex network of overlapping and interpenetrating acoustic spaces as observed by birds and animals will again have greater significance for the human community also.

Birds may be distinguished by the sounds of their flight. The great slow clapping of the eagle's wing is different indeed from the tremulous shaking of the sparrow against the air. "In reality I did not see the birds, but I heard the fast whir of their wings," wrote Frederick Philip Grove after crossing the Canadian prairies at night. The startled exodus of a flock of geese on a northern Canadian lake—a brilliant slapping of wings on water —is a sound as firmly imprinted in the mind of those who have heard it as any moment in Beethoven.

Some birds have furtive wings: "The owl's flight is too silent, its wing is down-padded. You may hear its beautiful call, but you will not hear its flight, even though it circle right around your head in the dusk." Only those who live close to the land can distinguish birds by the sounds of their wings in flight. Urban man has retained this facility only for insects and aircraft.

One notes with sadness how modern man is losing even the names of the birds. "I heard a bird" is a frequent reply I receive following a listening walk in a city.

"What bird?"

"I don't know." Linguistic accuracy is not merely a matter of lexicography. We perceive only what we can name. In a man-dominated world, when the name of a thing dies, it is dismissed from society, and its very existence may be imperiled.

Insects   The most easily recognized insect sounds for modern man are the most irritating. The mosquito, the fly and the wasp are easily distinguishable. The attentive listener can even tell the difference between male and female mosquitoes, the male usually sounding at a higher pitch. But only a specialist, such as a beekeeper, knows how to distinguish all the variants of the bee sound. Leo Tolstoy kept bees on his estate, and their sound is described in both *Anna Karenina* and *War and Peace*. "His ears were filled with the incessant hum in various notes, now the busy hum of the working bee flying quickly off, then the blaring of the lazy drone, and the excited buzz of the bees on guard protecting their property from the enemy and preparing to sting." When a queenless hive is dying, the beekeeper knows this too from the sound.

> The flight of the bees is not as in living hives, the smell and the sound that meet the beekeeper are changed. When the beekeeper strikes the wall of the sick hive, instead of the instant, unanimous response, the buzzing of tens of thousands of bees menacingly arching their backs, and by the rapid stroke of their wings making that whirring, living sound, he is greeted by a disconnected, droning hum from different parts of the deserted hive. . . . Around the entrance there is now no throng of guards, arching their backs and trumpeting the menace, ready to die in its defence. There is heard no more the low, even hum, the buzz of toil, like the singing of boiling water, but the broken, discordant uproar of disorder comes forth.

In his *Georgics* Virgil describes how the Roman beekeeper would "make a tinkling noise" with cymbals to attract the bees to a hive. He also describes vividly how two nests of bees would occasionally make war on one another with "a cry that is like the abrupt blasts of a trumpet."

The sounds of insects are produced in a surprising number of ways. Some, such as those of the mosquito and the drone bee, result from wing vibrations alone. The general range of wing frequencies in insects is between 4 and upward of 1,100 beats per second, and much of the pitched sound we hear from insects is produced by these oscillations. But when the butterfly moves its wings at between 5 and 10 times per second, the result is too faint and too low to be registered. In the honeybee, the wing beat frequency is 200 to 250 cycles per second and the mosquito (*Andes cantans*) has been measured at up to 587 cycles per second (c.p.s.). These frequencies would thus be the fundamentals of the resulting sounds, but as a rich spectrum of harmonics is also often present, the result may be a blurred noise with little discernible sensation of pitch.

Another type of sound produced by some insects is that created by tapping the ground. Such is the case in several species of termites. Large numbers of termites may hammer the ground in unison, presumably as a warning device, at a rate of about ten times a second, producing a faint drumming noise. Julian Huxley writes: "I remember waking up at night in camp, near Lake Edward, in the Belgian Congo, and hearing a strange clicking or ticking sound. A flashlight revealed that this was emanating from a column of termites which was crossing the floor of the tent under cover of darkness."

Still other insects, such as crickets and certain ants, produce stridulating effects by drawing parts of the anatomy called scrapers across other parts called files. The result of this filing activity is a complex sound, rich in harmonics. The variety of these stridulatory mechanisms is enormous, and by far the greatest number and variety of sounds produced by insects are produced in this manner.

Among the loudest of insects are the cicadas. They produce sound by means of ridged membranes or tymbals of parchment-like texture, close to the junction of the thorax and abdomen, which are set in motion by a powerful muscle attached to the inner surface; this mechanism produces a series of clicks in the same manner as does a tin lid when pressed in by the finger. The movement of the tymbals (amounting to a frequency of about 4,500 c.p.s.) is greatly amplified by the air chamber that makes up the bulk of the abdomen, so that the sound has been heard as far as half a mile away. In countries such as Australia and New Zealand, they create an almost oppressive noise when in season (December to March), though during the night they give way to the more gentle warbling of the crickets.

It is difficult to describe cicadas to one who does not know them, and when the young Alexander Pope was translating Virgil's line *sole sub ardenti resonant arbusta cicadis* (while the orchards echo to the harsh cicadas' notes and mine), he fell on the expedient for his English readers of communicating the same idea by means of a more recognizable sound: "The bleating sheep with my complaints agree."

Classical literature is full of references to cicadas as is oriental literature. They occur in the *Iliad* (where the Greek word *tettix*, τεττιξ, is often wrongly translated as "grasshopper") and in the works of Hesiod. Theocritus says that the Greeks kept them in cages for their singing ability, and this practice is still common among the children of southern lands. In *Phaedrus,* Plato has Socrates tell how the cicadas were originally men who were touched by the muses so that they devoted their lives to singing and, forgetting to eat, died to be reborn as insects. In Taoism, cicadas became associated with *hsien,* the soul, and images of cicadas are employed when preparing a corpse for burial, to assist the soul in disengaging itself from the body after death. The importance of the cicada in the soundscape of the South, as well as the symbolism it has provoked, has been overlooked since the comparatively recent northern drift of European and American civilization.

When they become part of the farmer's calendar, insects, like birds, arise out of the ambient soundscape to become signals for action: "May the fallows be worked for seed-time while the cicada overhead, watching the shepherds in the sun, makes music in the foliage of the trees."

The sounds of insects thus form rhythms, both circadian and seasonal, but entomologists have so far not measured these in sufficient detail for the soundscape researcher to be able to derive clear sound patterns from them. Difficulties have also been encountered in the analysis of the precise intensities and frequencies of insect sounds. This is because individual specimens are hard to isolate for recording purposes and also because the sounds insects make are generally complex frequency structures or broad-band noises, with harmonics often rising into the ultrasonic range. The locust *Schistocera gregaria* emits a sound of about 25 decibels when recorded very near the source, but the wing beat noise rises to 50 decibels when in flight. The flight noise of the desert locust has been measured as high as 67 decibels at a distance of 10 centimeters from the microphone. The sound output of many moths may be as little as 20 decibels quite near the source; while insects with hard wings and bodies, such as flies, bees and beetles, produce sounds up to 50 or 60 decibels. Since the human ear is more sensitive to sounds in the middle and upper frequency areas, insect sounds in the upper range (an average might be 400 to 1,000 c.p.s.) sound louder to the ear; but no human ear can hear the higher frequencies of the locust's call, which have been found to contain frequencies of 90,000 c.p.s.—that is, about two octaves higher than the human ear can detect.

For our purposes, however, one general feature of insect sounds is of interest. More perhaps than any other sound in nature, they give the impression of being steady-state or flat-line sounds. In part this may be an illusion, for many insect sounds are pulse modulated or varied in other subtle ways, but despite the "grainy" effect such modulations create, the impression with many insects is of a continuous, unvarying monotony. Like the straight line in space, the flat line in sound rarely occurs in nature, and we will not encounter it again until the Industrial Revolution introduces the modern engine.

The Sounds of Water Creatures   The sounds of the living are uttered only within a thin shell around the earth's surface—much less than 1 percent of its radius in width. They are confined to the land surface, the sea a few score fathoms below the surface and the air immediately above. But within this relatively small area the diversity of sounds produced by living organisms is bewilderingly complex. It is not our purpose here to survey all the sounds of nature and we will only touch on a few of the more unusual.

While many fish have no sound-producing mechanisms and no developed organs to hear sounds, many do produce unique sounds and some of these are very loud. Some fish, like sunfish or certain kinds of mackerel, make sounds by grinding or snapping their teeth. Others make sounds by expelling gas or by vibrating the gas bladder. One fish, the *Misgurnus,* makes a loudish noise by gulping air bubbles, and expelling them forcibly through its anus. At least thirty-four genera of fish produce sound by vibrating the gas bladder.

The songs of whales have been a subject of considerable recent study and some recordings of the humpback whale were produced commercially in 1970. The immediate and spectacular attention they received was partially attributable to the poignancy that the singers were an endangered species; but more than this, the songs were hauntingly beautiful. They also introduced many people, who had forgotten that the fish were their ancestors, to the echoing vaults of the ocean depths and united the feedback effects of popular electronic and guitar music with the multiple echoes of submarine acoustics—a subject to which we will return later. The songs of the humpback whale can be analyzed in musical terms. Each song seems to consist of a series of variations on constant themes or motifs, repeated differing numbers of times. Researchers are beginning to wonder if different herds or family groups of humpback whales may have different dialects.

Several of the crustaceans make sounds. The mantis shrimp (*Chloridella*) makes a loud noise by rubbing parts of its tail together, while the Florida spiny lobster makes a squeaking sound by rubbing a special flap on its antennae. Other crustaceans produce snapping, buzzing, hissing or even growling sounds which can often be heard on the seashore.

In the early spring, marshes in many parts of the world are filled with the sounds of frogs and toads. North America possesses a whole orchestra of performers: the narrow-mouthed toad bleats, the barking frog barks, the spring peeper chirps, the swamp cricket frog and the American toad trill, the least swamp cricket frog tinkles like an insect, the meadow frog rattles, the gopher frog snores, the green frog plays the banjo while the southern bullfrog belches.

When Julian Huxley visited America and heard the call of a bullfrog for the first time, he "refused to believe that it could proceed from a mere frog: it suggested a large and rather dangerous mammal, so loud it was and so low-pitched." Frogs are to North Americans what cicadas are to the Japanese or the Australians. The high resonant stridulations of some species such as the southern toad (*Bufo terrestris*) do indeed resemble cicadas, and sustained trills of the western toad (*Bufo cognatus*) have been recorded lasting as long as 33 seconds. But with the passing of the night, ardor wanes in the swamps; the voice of the bullfrog drops in pitch and the other instrumentalists gradually fade away.

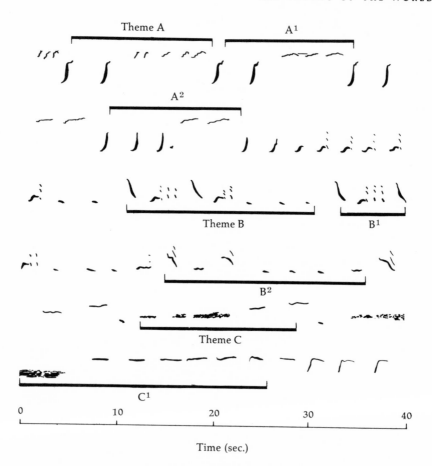

The song of the humpback whale, consisting of distinct themes and variations.

The Sounds of Animals

It is impossible to survey all the sounds produced by animals. I will mention only a few on our way to man himself. The carnivores produce the greatest range of individual sounds among animals, and many of these sounds, such as the roaring of lions, the howling of wolves or the laughing of hyenas, have such striking qualities that they impress themselves instantaneously on the human imagination. They present intense acoustic images. One hearing and they will never be mistaken or forgotten in a lifetime. They are among the great sounds that make history. Men who have heard tell of them only from the lips of the bard will still shudder at the thought of them.

Ludwig Koch recorded at least six distinct types of vocal expression in lions. Cubs yelp to obtain the attention of their parents and apparently yelp differently according to which parent they are soliciting. The maternal response is a rumbling sound with a certain grunting quality. There is a "pleasure call," chiefly noticed among lions in captivity, which is initiated by the appearance of the keeper. The feeding sound, when the beast is alone and undisturbed, is a deep, gentle growl. Just when the prey is seized, lions produce a short and terrifying bark of ferocity. Finally there is the true roaring, usually heard at night, rarely heard during daylight. When roaring, lions will sometimes set their mouths close to the ground to assist the resonance and reverberation of their voices.

Lions do not purr. But leopards do and so do cheetahs, loudly. Aside from the hissing and spitting noises that most cats produce when angry, each cat has a repertoire of unique sounds. For instance, the puma has a loud wailing scream which Julian Huxley says could "be mistaken for that of a child," and the cubs produce a whistling note. Tigers are less noisy animals than lions but they do have a crazy love call, like that of ordinary cats, but greatly magnified.

The howling of wolves is haunting and isolating. Usually the leader of the pack will begin in solo; then the others will join in chorus, howling first, then descending to a ragged yelping bark. In the wolf call we encounter a vocal ritual which defines the territorial claim of the pack to an acoustic space—in just the same way as the hunter's horn lays claim to the forest or the church bell to the parish.

The sounds made by primates have always interested and amused man. They exist in great diversity, varying from whistling, screaming and chattering to grunting and roaring. Some are very loud. The howler monkey of South America has the strongest voice for its size of any mammal, and it is said to carry nearly five kilometers in open country and three kilometers in dense forest. The animal has a special bellows-like structure in its larynx to assist it in producing such a volume of sound. So far no exact measurements of these animals have been made. We measured Hoolack gibbons at a peak level of 110 dBA* outside their cages in the Vancouver Zoo. Julian Huxley tells of a friend having heard gibbons in the London Zoo from Oxford Circus, during quiet early morning hours. That would be a distance of nearly two kilometers.

The gorilla is the only primate to have discovered a nonvocal sound mechanism: it drums on its chest with its fist, producing a loud, hollow sound. This is done both when making vocal sounds and on its own. The gorilla has discovered the property of resonance, independent of the natural mechanism of the voice box. It seems forever on the verge of discovering the musical instrument without being able to make the transition from personal to artificial sound. So far as we know only man has done this.

---

*Decibels are more accurately designated by the addition of A, B or C to their abbreviation of dB. DBA indicates that the lower frequencies of the sound are discriminated against by a weighting network in the measuring instrument in a manner roughly equivalent to the human ear's discrimination against low-frequency sounds. DBB indicates less of such discrimination, while dBC represents nearly flat response to the sound being measured.

Man Echoes the Soundscape in Speech and Music

All the animal sounds mentioned in these pages fall into a few general categories. They may be either sounds of warning, mating calls, exchanges between mother and offspring, food sounds or social sounds. All of these are identifiable also in the vocal utterances of man, and the purpose of the remainder of this book will be to illustrate how they have been worked out in human communities throughout history.

To begin this we should draw attention to the fact that many of the signals communicated among animals—those of hunting, warning, fright, anger or mating—often correspond very closely in duration, intensity and inflection to many human expletives. Man also may growl, howl, whimper, grunt, roar and scream. This, together with the fact that man often shares the same geographic territories with the animals, obviously accounts for their frequent appearance in his folklore and rituals. In these rituals, such as the monkey dance of the Balinese, the voices of the animals are conjured by man in striking imitations. Marius Schneider writes:

> One must have heard them to realize how extremely realistically aboriginals are able to imitate animal noises and the sounds of nature. They even hold "nature concerts" in which each singer imitates a particular sound (waves, wind, groaning trees, cries of frightened animals), "concerts" of surprising magnificence and beauty.

We are at that remote time in prehistory when the double miracle of speech and music occurred. How did these activities come about? It would be rash to insist that speech originated exclusively in the onomatopoeic mimicry of the natural soundscape. But that the tongue danced and still continues to dance with the soundscape, there can be no doubt. Poets and musicians have kept the memory alive, even if modern man has acquiesced into bespectacled muttering. Concerning the flattening out of human vocal style, the linguist Otto Jespersen has written:

> Now, it is a consequence of advancing civilization that passion, or, at least, the expression of passion, is moderated, and we must therefore conclude that the speech of uncivilized and primitive man was more passionately agitated than ours, more like music or song. . . . Although we now regard the communication of thought as the main object of speaking. . . it is perfectly possible that speech has developed from something which had no other purpose than that of exercising the muscles of the mouth and throat and of amusing oneself and others by the production of pleasant or possibly only strange sounds.

Onomatopoeia mirrors the soundscape. Even with our advanced speech today we continue, in descriptive vocabulary, to cast back sounds

heard in the acoustic environment; and it may even be that the more
sophisticated acoustic extensions of man—his tools and signaling devices
—also continue, to some extent, to extend the same archetypal patternings.
We have been discussing animals. Among the characters of his speech,
man has numerous words to describe the animal sounds which are nearest
to him. These are verbs, action words, and most of them are onomatopoeic
still:

> a dog *barks*
> a puppy *yelps*
> a cat *meows* and *purrs*
> a cow *moos*
> a lion *roars*
> a goat *bleats*
> a tiger *snarls*
> a wolf *howls*
> a mouse *squeaks*
> a donkey *brays*
> a pig *grunts* or *squeals*
> a horse *whinnies* or *neighs*

The English language reproduces only those animals with which English
man, in his many migrations, has found himself in closest contact. But the
English language knows no special words for those animals remote from
English man: the galago, the mangabey, the llama or the tapir.

Some day a linguist ought to investigate those even more primeval
human imitations, still found in folklore or children's rhymes, where we
have a decisive attempt to duplicate the actual sounds of animals and birds.
The differences between languages are interesting.

> Dog: bow-wow (E), arf-arf (A), gnaf-gnaf (F), how-how (Ar), gaû-gaû
> (V), won-won (J), kwee-kwee (L).
> Cat: purr-purr (E), ron-ron (F), schnurr-schnurr (G).
> Sheep: baa-baa (E), méé-méé (Gr, J, M), maa'-maa' (Ar).
> Bee: buzz (E), zŭz-zŭz (Ar), bun-bun (J), vū-vū (V).
> Cockerel: cock-a-doodle-doo (E), cock-a-diddle-dow (Shakespeare),
> kikeriki (G), kokke-kokkō (J), kiokio (L).*

To this list one could add many other interesting words, for example,
sneeze: kerchoo (A), atishoo (E), achum (Ar), cheenk (U), kakchun (J),
ach-shi (V).

Such imitations are limited, of course, to the phonemes available for
their reproduction in any given language; but such a study, if pursued

---

*Abbreviations for languages are: E—English, A—American, F—French, Ar—Arabic,
V—Vietnamese, J—Japanese, G—German, Gr—Greek, M—Malay, U—Urdu, L—Lokele tribe
of the Congo.

diligently enough, might bring us closer to measuring how the critical features of natural sounds are perceived by different peoples.

In onomatopoeic vocabulary, man unites himself with the soundscape about him, echoing back its elements. The impression is taken in; the expression is thrown back in return. But the soundscape is far too complex for human speech to duplicate, and so it is in music alone that man finds that true harmony of the inner and outer world. It will be in music too that he will create his most perfect models of the ideal soundscape of the imagination.

THREE

~~~~~~

The Rural Soundscape

The Hi-Fi Soundscape   In discussing the transition from the rural to the urban soundscape, I will be using two terms: hi-fi and lo-fi. They need to be explained. A hi-fi system is one possessing a favorable signal-to-noise ratio. The hi-fi soundscape is one in which discrete sounds can be heard clearly because of the low ambient noise level. The country is generally more hi-fi than the city; night more than day; ancient times more than modern. In the hi-fi soundscape, sounds overlap less frequently; there is perspective—foreground and background: ". . . the sound of a pail on the lip of a well, and the crack of a whip in the distance"—the image is Alain-Fournier's to describe the economic acoustics of the French countryside.

The quiet ambiance of the hi-fi soundscape allows the listener to hear farther into the distance just as the countryside exercises long-range viewing. The city abbreviates this facility for distant hearing (and seeing) marking one of the more important changes in the history of perception.

In a lo-fi soundscape individual acoustic signals are obscured in an overdense population of sounds. The pellucid sound—a footstep in the snow, a church bell across the valley or an animal scurrying in the brush —is masked by broad-band noise. Perspective is lost. On a downtown street corner of the modern city there is no distance; there is only presence. There is cross-talk on all the channels, and in order for the most ordinary sounds to be heard they have to be increasingly amplified. The transition from the hi-fi to the lo-fi soundscape has taken place gradually over many centuries and it will be the purpose of several of the following chapters to measure how it has come about.

In the quiet ambiance of the hi-fi soundscape even the slightest disturbance can communicate vital or interesting information: "He was disturbed in his meditation by a grating noise from the coachhouse. It was the vane on the roof turning round, and this change in the wind was the signal for a disastrous rain." The human ear is alert, like that of an animal. In the stillness of the night a paralyzed old lady in a story by Turgenev can hear the moles burrowing underground. "That's when it's good," she reflects; "no need to think." But poets do think about such sounds. Goethe, his ear pressed to the grass: "When I hear the humming of the little world among the stalks, and am near the countless indescribable forms of the worms and insects, then I feel the presence of the Almighty, Who created us in his own image. . . ."

From the nearest details to the most distant horizon, the ears operated with seismographic delicacy. When men lived mostly in isolation or in small communities, sounds were uncrowded, surrounded by pools of stillness, and the shepherd, the woodsman and the farmer knew how to read them as clues to changes in the environment.

Sounds of the Pasture   The pasture was generally quieter than the farm. Virgil describes it well:

> . . . Hyblaean bees coax you with a gentle humming through the gates of sleep . . . you will have the vine-dresser singing to the breezes, while all the time your dear full-throated pigeons will be heard, and the turtle-dove high in the elm will never bring her cooing to an end.

Shepherds may, as Lucretius suggests, have got the hint of singing and whistling from the sound of the wind. Or it may have been from the birds. Virgil says that Pan taught the shepherd "how to join a set of reeds with wax" as a means of conversing with the landscape.

> Sweet is the whispered music of yonder
>   pinetree by the springs,
> goatherd, and sweet too thy piping. . . .
>   Sweeter, shepherd, falls thy song
> than yonder stream that tumbles
>   plashing from the rocks.

Shepherds piped and sang to one another to while away the lonely hours, as the dialogue form of Theocritus's *Idylls* and Virgil's *Eclogues* shows us; and the delicate music of their songs forms perhaps the first and certainly the most persistent of the man-made sonic archetypes. Centuries of piping have produced a referential sound that still suggests the serenity of the pastoral landscape clearly, though many traditional literary images and devices are beginning to slip away. The solo woodwind always paints the pastorale, and this archetype is so suggestive that even such a grandiloquent orchestrator as Berlioz slims his orchestra to a duet between solo English horn and oboe to draw us gently into the country (*Symphonie Fantastique,* third movement).

In the still landscape of the country, the clear dulcet tones of the shepherd's pipe took on miraculous powers. Nature listened, then responded sympathetically: "The music struck the valleys and the valleys tossed it to the stars—till the lads were warned to drive home and to count their sheep, by Vesper, as he trod unwelcome into the listening sky." Theocritus was the first poet to make the landscape echo the sentiments of the shepherd's pipes, and pastoral poets have been copying him ever since.

> ... Practice country songs on a light shepherd's pipe ...
> teaching the woods to echo back the charms of Amaryllis

says Virgil. For a recurrence of this miraculous power in music, we must wait until the nineteenth-century romanticists.

The pastoral soundscape our poets have been describing continued on into the nineteenth century. Alain-Fournier describes it in France: "Now and then the distant voice of a shepherdess, or of a boy calling to a companion from one clump of firs to another, had risen in the great calm of the frozen afternoon." The juncture between town and pasture is attractively captured in this description by Thomas Hardy:

> The shepherd on the east hill could shout out lambing intelligence to the shepherd on the west hill, over the intervening town chimneys, without great inconvenience to his voice, so nearly did the steep pastures encroach upon the burghers' backyards. And at night it was possible to stand in the very midst of the town and hear from their native paddocks on the lower levels of greensward the mild lowing of the farmer's heifers, and the profound, warm blowings of breath in which those creatures indulge.

Sounds of the Hunt   A quite different type of sonic archetype has come down to us from the hunt, for the horn transpierces the gloom of the forest wilderness with heroic and bellicose tones. Almost all cultures seem to have employed some type of horn in association with warfare and the hunt. The Romans used a hooped horn of conical tube as a signaling instrument for their armies and there are numerous references to it in Dion, Ovid and Juvenal; but when Rome declined, the art of smelting brass seems to have disappeared and with it went a special sound. When "Sigmund blew the horn that had been his father's, and urged on his men," it was an animal's horn that he blew. The same type appears in the pages of the *Chanson de Roland.* But by the fourteenth century, the skill of smelting brass had been recovered and brilliant metallic tones began to echo across Europe.

By the sixteenth century the *cor de chasse* had taken on something like a definitive character and it is this instrument that gained special significance in the European soundscape, a significance that has lasted until recent times. In the days when hunting was popular, the countryside can scarcely have ever been free of horn calls, and the elaborate code of signals must have been widely known and understood.

As the *cor de chasse* was an open horn possessing only a few natural harmonics, its various signals possessed more distinctive rhythmic than melodic character. The various codes that have been preserved are of considerable complexity and, of course, vary greatly from country to country. They may be classified as follows:

1. brief calls intended to cheer on the hounds, to give warning, to call for aid or to indicate the circumstances of the hunt;
2. a special fanfare for each animal (several for the stag, depending on his size and antlers);
3. fancy tunes to begin or close a hunt, or sounded as special signs of joy.

Tolstoy has given us a good account of the festive nature of the hunt in Russia.

The hounds' cry was followed by the bass note of the hunting cry for a wolf sounded on Danilo's horn. The pack joined the first three dogs, and the voices of the hounds could be heard in full cry with the peculiar note which serves to betoken that they are after a wolf. The whippers-in were not now hallooing, but urging on the hounds with cries of "Loo! loo! loo!" and above all the voices rose the voice of Danilo, passing from a deep note to piercing shrillness. Danilo's voice seemed to fill the whole forest, to pierce beyond it, and echo far away in the open country.

A contemporary recollection by a young woman shows how strong the heritage of the hunt still is in northern Germany.

It was still quite dark when one of the hunters ceremonially opened the hunt with a fanfare on his horn. Unless the stretch of land to be hunted was an open field, the only method of communication between hunters and beaters was through horn signals. During the formation, in which the hunters enclosed the area on three sides and the beaters on one, everyone was very quiet so as not to disturb the animals. The silence was broken by a horn signal, which was answered by a terrible, shrill toot from a single-pitch trumpet (which looked like a toy trumpet) blown by one of the beaters. We started to attack the land in front of us with rattles, pots, pans, noisemakers of every sort and shouts in all modulations. Frightened by the noise, every living creature was stirred up and chased out of its shelter in the direction of the hunters. As children we just loved to make the loudest noises. . . .

At the end of the day, everybody gathered around and listened to the horn player blowing the fanfares for the dead animals. There was a signal for every animal, and I remember that the one for the fox was the most beautiful, whereas the one for the rabbit was quite short and simple. At the end of the day, in the evening darkness, the hunt was ended by a cheerful, almost triumphant fanfare.

The hunting horn presents us with a sound of great semantic richness. On one level its signals provide a code which all participants understand. On another level it takes on a symbolical significance, suggesting free spaces and the natural life of the country. I also spoke of the hunting horn as an archetypal sound. Only sound symbols which are carried forward century after century qualify for this distinction, for they knit us with ancient ancestral heritages, providing continuity at the deepest levels of consciousness.

The Post Horn   Another sound of similar character which was also ubiquitous on the European scene was the post horn. It too persisted for centuries, for it began in the sixteenth century when the administration of the post was taken over by the family of Thurn and Taxis, and as the postal routes extended from Norway to Spain so did the horn calls (Cervantes mentions them). In Germany the last post horns were heard in 1925. In England the post horn was still in use in 1914 when the London-to-Oxford mail was conveyed by road on Sundays. In Austria, horns were also heard until after the First World War, and even today no one is permitted to carry or sound a post horn, thereby enhancing the sentimental symbolism of the instrument (Article 24 of the Austrian Postal Regulations, 1957).

The post horn also employed a precise code of signals to indicate different types of mail (express, normal, local, packages) as well as calls for arrival, departure and distress, and indications for the number of carriages and horses—in order that the changing stations might receive advance warning. In Austria a recruit was given six months to learn the signals and if he failed, he was dismissed.

Through the narrow streets and across the country landscape the post horn was heard, in the villages and the alleys of cities, at the gates of castles above and by the monasteries below in the valleys—everywhere its echo was known, everywhere it was greeted joyfully. It touched all the strings of the human heart: hope, fear, longing and homesickness—it awakened all feelings with its magic.

Thus the symbolism of the post horn worked differently from that of the hunting horn. It did not draw the listener out into the landscape but, working in reverse, brought news from far away to home. It was centripetal rather than centrifugal in character and its tones were never more pleasant than when it approached the town and delivered its letters and parcels to the expectant.

Sounds of the Farm   By comparison with the quiet life of the
pasture and the shrill celebrations of the hunt, the soundscape of the farm
provides a general turmoil of activities. Each of the animals has its own
rhythm of sound and silence, of arousal and repose. The cock is the eternal
alarm clock, and the dog's bark is the original telegraph—for one learns
when acreage has been invaded by a stranger from the dog's barking,
passed from farm to farm.

Many of the sounds of the farm are heavyweight, like the slow,
tramping hooves of cattle and draft horses. The farmer's feet, too, move
slowly. Virgil tells us of "ponderous-moving wagons," of threshers and
"the immoderate weight of the harrow." He also gives us an interesting
acoustic picture of the Italian farmhouse after dark.

> One farmer stays awake and splits up wood
> For torches with his knife. And all the while
> His wife relieves her lengthy task with song,
> And runs the squeaky shuttle through the warp,
> Or boils down sweetened wine-must over flame,
> And skims with leaves the bubbling cauldron's wave.

Some of the sounds of the farm have changed little over the centuries,
particularly those suggesting the commotion of heavy work; and the voices
of animals too have given a consistency of tone to the farm soundscape.
But there are also vernaculars. From my own youth I recall a few. The first
that comes to mind is the churning of butter. As the churn was pumped
for half an hour or more, an almost imperceptible change in tone and
texture occurred as the slopping cream gradually turned to butter. The
hand-operated pump, also on the decline, now snaps into memory as a
soundmark of my youth, though at the time I listened to it carelessly.
There were others too, like the ubiquitous cackling of geese, or the swoosh
and bang of the screen door. In the winter there was the heavy stamping
of snow boots in the front hall, or the scream of sleigh runners over
hard-packed country roads. In the silence of the winter night there might
be a sudden crack as a nail sprang from a board in the intense cold. And
there were the deep pedal tones that came again and again in the chimney
flue during night winds. Then there were the regular rhythms like the gong
which brought us in for dinner, or the whirring of the windmill, which the
women put in motion at four o'clock each day to pump water for the
returning cattle.

I have defined keynote as a regular sound underpinning other more
fugitive or novel sound events. The keynotes of the farm were numerous,
for farming is a life with little variation. Keynotes may influence the
behavior of the people or set up rhythms that are carried over into other
aspects of life. One example will have to suffice. In the Russia of Tolstoy,
the peasants kept whetstones in little tin boxes strapped to their waists,

and the rhythmic rattling of these boxes formed a vernacular keynote during the haying months.

> The grass cut with a juicy sound, and was at once laid in high, fragrant rows. The mowers from all sides, brought closer together in the short row, kept urging one another on to the sound of rattling tin boxes and clanging scythes, and the hiss of the whetstones sharpening them, and happy shouts.

Returning from the fields, the rhythms of the day's work were extended into song.

> The peasant women, with their rakes on their shoulders, gay with bright flowers, and chattering with ringing, merry voices, walked behind the cart. One wild untrained female voice broke into a song, and sang it alone through a verse, and then the same verse was taken up and repeated by half a hundred strong, healthy voices of all sorts, coarse and fine, singing in unison . . . the whole meadow and distant fields all seemed to be shaking and singing to the measures of this wild, merry song with its shouts and whistles and clapping.

Russia is, of course, not the only place where the rhythms of work have been carved into folk song, but folk song suggested by work always carries a heavy stress. This becomes clear if we compare the music of the farm laborer with the levity of the shepherd's pipes. I do not think it would be going too far to suggest that man only discovers lilt and lyricism in music to the extent that he frees himself from physical labor.

Noise in the Rural Soundscape
The rural soundscape was quiet, but it experienced two profound acoustic interruptions: the noise of war and the "noise" of religion.

Virgil, whose life was frequently interrupted by the Roman wars, laments these intrusions into the pastoral life.

> Such was the life that golden Saturn
> lived upon earth:
> Mankind had not yet heard the bugle
> bellow for war,
> Nor yet heard the clank of the sword
> on the hard anvil. . . .

To Virgil the sounds of war were brass and iron, and the acoustic image remains intact to this day, though to it must be added the explosions of gunpowder from the fourteenth century onward.

The world's literature is full of battles. Poets and chroniclers seem always to have been amazed at the noise they made. The Persian epic poet Ferdowsi is typical.

> At the shouts of the Divs and the noise made by the black dust rising, the thunder of drums and the neighing of war-horses, the mountains were rent and the earth cleft asunder. So fierce a combat had been seen

by no man before. Loud was the clash of the battle-axes and the clatter of swords and of arrows; the warriors' blood turned the plain into marsh, the earth resembled a sea of pitch whose waves were formed of swords, axes and arrows.

Armies decorated for battle presented a visual spectacle, but the battle itself was acoustic. To the actual noise of clashing metal, each army added its battlecries and drumming in an attempt to frighten the enemy. Noise was a deliberate military stratagem; the ancient Greek generals advocated it: "One should send the army into battle shouting, and sometimes on the run, because their appearance and shouts and the clash of arms confound the hearts of the enemy." From Tacitus comes an interesting description of a German war chant called *baritus:*

> By the rendering of this they not only kindle their courage, but, merely by listening to the sound, they can forecast the issue of an approaching engagement. For they either terrify their foes or themselves become frightened, according to the character of the noise they make upon the battlefield; and they regard it not merely as so many voices chanting together but as a unison of valour. What they particularly aim at is a harshly intermittent roar; and they hold their shields in front of their mouths, so that the sound is amplified into a deeper crescendo by the reverberation.

When the Moors attacked the Castilians in 1085, they employed African drummers who, according to the *Poema del Cid,* had never before been heard in Europe. The noise terrified the Christians but "the good Cid Campeador" pacified his army, promising to capture the drums and deliver them to the Church. The association of noise with both warfare and religion was not fortuitous, and we shall frequently find reason throughout this book for coupling them together. Both activities are eschatological, and undoubtedly an awareness of this fact lies behind the peculiar bending of the Latin word *bellum* (war) into the Low German and Old English *bell(e)* (meaning "to make a loud noise") before its final imprint on the instrument which gave Christianity its acoustic signal.

One further example will reinforce the relationship between religion, warfare and noise, for it is a description of a religious battle which seems to have been fought by sound alone.

> It was at three o'clock on August 14th, 1431, that the crusaders, who were encamped in the plain between Domazlice and Horsuv Tyn, received the news that the Hussites, under the leadership of Prokop the Great, were approaching. Though the Bohemians were still four miles off, the rattle of their war-wagons and the song, "All ye warriors of God," which their whole host was chanting, could already be heard. The enthusiasm of the crusaders evaporated with astounding rapidity.... The German camp was in utter confusion. Horsemen were streaming off in every direction, and the clatter of empty wagons

being driven off almost drowned the sound of that terrible singing. . . .
So ended the Bohemian crusade.

The point I am trying to make with the diverse descriptions of these
pages is that while the original soundscape was generally quiet, it was
deliberately punctuated by the aberrational noises of war. The *other* occasion for loud noise was religious celebration. It was then that the rattles
and drums and sacred bones were brought out and sounded vigorously to
produce what for elementary man was certainly the biggest acoustic event
of civil life. There is no doubt that these activities were a direct imitation
of the frightening sounds of nature already studied, for they too had divine
origins. Thunder was created by Thor or Zeus, storms were divine combats,
cataclysms were divine punishments. We recall that the word of God
originally came to man through the ear, not the eye. By gathering his
instruments and making an impressive noise, man hoped in his turn to
catch the ear of God.

Sacred Noise and Secular Silence   Throughout the several
hundred pages of his *Mythologiques II,* the anthropologist Lévi-Strauss has
developed an argument for placing noise in parallel with the sacred and
silence in the same relationship with the profane.* The Lévi-Strauss argument, regarded from the vantage point of the modern noise-riddled world,
may appear obscure; but soundscape studies help to clarify it. The profane
world was, if not silent, quiet. And if we think of "noise" in its less
pejorative sense as any big sound, the coupling of noise and sacred is easier
to interpret.

Throughout this book we are going to discover that a certain type of
noise, which we may now call Sacred Noise, was not only absent from the
lists of proscripted sounds which societies from time to time drew up, but
was, in fact, quite deliberately invoked as a break from the tedium of
tranquility. Samuel Rosen confirmed this when he studied the acoustic
climate of a quiet tribal village in the Sudan.

> In general, the sound level in the villages is below 40 db on the C scale
> of the sound level meter except occasionally at sunrise or soon there-
> after when a domestic animal such as a rooster, lamb, cow or dove
> makes itself heard. During six months of the year, heavy rains occur
> about three times a week with one or two loud claps of thunder. A
> few men engage in some productive activities such as beating palm
> fronds with a wooden club. But the absence of hard reverberating
> surfaces, such as walls, ceilings, floors and hard furniture, etc., in the

---

*I must warn the reader that Lévi-Strauss informs me that the Sacred Noise theory
developed in this book bears "little relationship, if any" to what he has written. Nevertheless,
I must give him credit for igniting my imagination.

vicinity apparently accounts for the low intensity levels measured on the sound level meter: 73–74 db at the worker's ear.

The loudest noises (over 100 decibels) were encountered when the villagers were singing and dancing, which occurred for the most part "over a two-month period celebrating the spring harvest" (i.e. a religious festival).

Throughout Christendom the divine was signaled by the church bell. It is a later development of the same clamorous urge, which had earlier been expressed in chanting and rattling. The interior of the church, too, reverberated with the most spectacular acoustic events, for to this place man brought not only his voice, raised in song, but also the loudest machine he had till then produced—the organ. And it was all designed to make the deity listen.

Aside from the spectacular celebrations of warfare and religion, rural and even town life was tranquil. There are many towns still, the world over, where life moves uneventfully, almost by stealth. Poor towns are quieter than prosperous towns. I have visited towns in Burgenland (Austria) where the only sound at midday is the flapping of storks in their chimney nests, or dusty towns in Iran where the only motion is the occasional swaying walk of a woman carrying water while the children sit mutely in the streets. Peasants and tribesmen the world over participate in a vast sharing of silence.

FOUR

~~~~~~

From Town to City

The two great turning points in human history were the change from nomadic to agrarian life, which occurred between ten and twelve thousand years ago, and the transition from rural to urban life, which has occupied the most recent centuries. As this later development has occurred, towns have grown into cities and cities have swollen out to cover much that was formerly rural land.

In terms of the soundscape, a practical division of developing urbanization is, as in so many other matters as well, the Industrial Revolution. In the present chapter I will consider only the pre-industrial period, leaving the sequel to be taken up in Part Two of the book. A proper consideration of pre-industrial town and city life would need a much more thorough treatment than it can be given here. Town and city life diverged greatly before the Industrial and Electric revolutions began to level it, but I can only hope to hint at some of the variations, while dealing particularly with the European scene. There is a practical reason for this limitation: the accessibility of documentation.

Looking at the profile of a medieval European city we at once note that the castle, the city wall and the church spire dominate the scene. In the modern city it is the high-rise apartment, the bank tower and the factory chimney which are the tallest structures. This tells us a good deal about the prominent social institutions of the two societies. In the soundscape also there are sounds that obtrude over the acoustic horizon: keynotes, signals and soundmarks; and these types of sounds must accordingly form the principal subject of our investigation.

Making God Listen   The most salient sound signal in the Christian community is the church bell. In a very real sense it defines the

53

community, for the parish is an acoustic space, circumscribed by the range of the church bell. The church bell is a centripetal sound; it attracts and unifies the community in a social sense, just as it draws man and God together. At times in the past it took on a centrifugal force as well, when it served to frighten away evil spirits.

Church bells appear to have been widespread in Europe by the eighth century. In England they were mentioned by The Venerable Bede at the close of the seventh century. Of their gigantic presence, Johan Huizinga writes in *The Waning of the Middle Ages:*

> One sound rose ceaselessly above the noises of busy life and lifted all things unto a sphere of order and serenity: the sound of bells. The bells were in daily life like good spirits, which by their familiar voices, now called upon the citizens to mourn and now to rejoice, now warned them of danger, now exhorted them to piety. They were known by their names: big Jacqueline, or the bell Roland. Every one knew the difference in meaning of the various ways of ringing. However continuous the ringing of the bells, people would seem not to have become blunted to the effect of their sound.
>
> Throughout the famous judicial duel between two citizens of Valenciennes, in 1455, the big bell, "which is hideous to hear," says Chastellain, never stopped ringing. What intoxication the pealing of the bells of all the churches, and of all the monasteries of Paris, must have produced, sounding from morning till evening, and even during the night, when a peace was concluded or a pope elected.

Throngs of pitched bells or carillons were especially popular in the Netherlands, where they irritated Charles Burney on his European tours. "The great convenience of this kind of music," Burney wrote, "is that it entertains the inhabitants of a whole town, without giving them the trouble of going to any particular spot to hear it." At a suitable distance, however, church bells could be powerfully evocative, for the strident noises of the clappers are lost and they are given a legato phrasing which wind currents or water will modulate dynamically, so that even a few simple and not very good bells can provide hours of pleasant listening. Perhaps no sound benefits more from distance and atmosphere. Church bells form a sound complement to distant hills, wrapped in blue-gray mist. Traveling a similar route to that of Charles Burney, yet keeping to the rivers and canals and avoiding the cities, Robert Louis Stevenson experienced church bells transformed in this way.

> On the other side of the valley a group of red roofs and a belfry showed among the foliage. Thence some inspired bell-ringer made the afternoon musical on a chime of bells. There was something very sweet and taking in the air he played; and we thought we had never heard bells speak so intelligibly, or sing so melodiously, as these. . . .

There is so often a threatening note, something blatant and metallic, in the voice of bells, that I believe we have fully more pain than pleasure from hearing them; but these, as they sounded abroad, now high, now low, now with a plaintive cadence that caught the ear like the burthen of a popular song, were always moderate and tunable, and seemed to fall in with the spirit of still, rustic places, like the noise of a waterfall or the babble of a rookery in spring.

Wherever the missionaries took Christianity, the church bell was soon to follow, acoustically demarking the civilization of the parish from the wilderness beyond its earshot.* The bell was an acoustic calendar, announcing festivals, births, deaths, marriages, fires and revolts. In Salzburg, from a small ancient hotel room, I listened to the innumerable bells ring slowly, just a shade slower than what one would expect, producing little tensions in the mind as anticipation fell a fraction of a second short of reality. And at San Miguel de Allende in Mexico I remember watching the convicts in the belfry, putting the giant bells into motion by tugging at their rims with heavy, awkward movements.

The Sound of Time   It was during the fourteenth century that the church bell was wedded to a technical invention of great significance for European civilization: the mechanical clock. Together they became the most inescapable signals of the soundscape, for like the church bell, and with even more merciless punctuality, the clock measures the passing of time audibly. In this way it differs from all previous means of telling time —water clocks, sand clocks and sundials—which were silent.

> The church clock struck eleven. The air was so empty of other sounds that the whirr of the clock-work immediately before the strokes was distinct, and so was also the click of the same at their close. The notes flew forth with the usual blind obtuseness of inanimate things— flapping and rebounding among walls, undulating against the scattered clouds, spreading through their interstices into unexplored miles of space.

The clock bell had a great advantage over the clock dial, for to see the dial one must face it, while the bell sends the sounds of time rolling out uniformly in all directions. No European town was without its many clocks.

> Other clocks struck eight from time to time—one gloomily from the gaol, another from the gable of an almshouse, with a preparative creak of machinery, more audible than the note of the bell; a row of tall,

*Typically, while both the Muslim and Christian faiths have important signaling devices, the Jewish faith, which is not missionizing, does not.

varnished case-clocks from the interior of a clock-maker's shop joined in one after another just as the shutters were enclosing them, like a row of actors delivering their final speeches before the fall of the curtain; then chimes were heard stammering out the Sicilian Mariners' Hymn; so that chronologists of the advanced school were appreciably on their way to the next hour before the whole business of the old one was satisfactorily wound up.

Clocks regulated the movements of the town with militant imperiousness. Occasionally they rose to the status of soundmarks. (How well I remember the erratic pentatonic descent of the clock bell in the Kremlin wall—the only whimsy about the place.) Affectionately regarded by the inhabitants, some old clocks are even specifically exempted from anti-noise legislation, as is the case with the post office clock in Brantford (Ontario).

The historian Oswald Spengler believed that it was the mechanical clock that gave Europe (and particularly Germany) its sense of historical destiny.

Amongst the Western peoples, it was the Germans who discovered the mechanical *clock*, the dread symbol of the flow of time, and the chimes of countless clock towers that echo day and night over West Europe are perhaps the most wonderful expression of which a historical world-feeling is capable.

The association of clocks and church bells was by no means fortuitous; for Christianity provided the rectilinear idea of the concept of time as progress, albeit spiritual progress, with a starting point (Creation), an indicator (Christ) and a fateful conclusion (the Apocalypse). Already in the seventh century it was decreed in a bull of Pope Sabinianus that the monastary bells should be rung seven times each day, and these punctuation points became known as the canonical hours. Time is always running out in the Christian system and the clock bell punctuates this fact. Its chimes are acoustic signals, but even at a subliminal level the incessant rhythm of its ticking forms a keynote of unavoidable significance in the life of Western Man. Clocks reach into the recesses of night to remind man of his mortality.

Other Focal Points   Clocks are centripetal sounds; they unify and regulate the community. But they are not the only centripetal sounds. From early times in agricultural territories, the mill was a prominent institution at the center of town life, and its sound was as familiar as the voices of the inhabitants themselves. In Ecclesiastes (12: 3–5) the author sketches a sinister soundscape when "the women grinding the meal cease to work ... when the noise of the mill is low, when the chirping of the sparrow grows faint and the songbirds fall silent." Water wheels used for milling were recorded in Rome as far back as the first century B.C., and while many other Roman arts disappeared, only to be rediscovered in the late Middle Ages, the water mill survived, for there are frequent references to it throughout early medieval literature.

Grinding grain was not the only work done by the mills, for by the early fourteenth century there were also papermills and sawmills. By then mills also turned grinding machines for the armorers and later they ran the hammering and cutting machines of ironworks. This is why so many towns were founded on the banks of rivers or streams, where water power was available.

> Where the lake became a brook, there were two or three mills. Their wheels seemed to run after each other, splashing water, like silly girls. I used to stay there long hours, watching them and throwing pebbles in the waterfalls to see them bounce and then fall again to disappear under the whirling round of the wheel. From the mills one could hear the noise of the millstones, the millers singing, children screaming, and always the squeaking of the chain over the hearth while the polenta was being stirred. I know this because the smoke coming out of the chimney always preceded the occurrence of this new, strident note in the universal concert. In front of the mills, there was a constant coming and going of sacks and flour-covered figures. Women from nearby villages came and chatted with the women of the mills while their grain was being ground. Meanwhile, the little donkeys, freed of their loads, greedily enjoyed the bran mash prepared as a treat for them on the occasion of the trips to the mills. When they finished, they started to bray, merrily stretching ears and legs. The miller's dog barked and ran around them with playful assaults and defenses. I tell you, it was indeed a very lively scene and I couldn't think of anything better.

To those living in the mill itself, life was never without the "patter" (Thomas Hardy's word) of the big wheel, to which the little ones mumbled responsively, producing "a remote resemblance to the stopped diapason in an organ." Later the mill, now equipped with a strident whistle, began to take on a more dominating aspect. We jump ahead momentarily to 1900, to a description of Dryomov, Russia, in the words of Maxim Gorky. "Awakening in the pearly gloom of an autumn dawn, Artamonov senior would hear the summoning blast of the mill whistle. Half an hour later would commence the indefatigable murmur and rustle, the accustomed, dull, but powerful din of labour." Another sound that continued all day within earshot of most of the residents of the early town was the blacksmith. ". . . the sounds could not have been more distinct if they had been dropped down a deep well. From the blacksmith shop . . . came a *tang-tang*. A bee droomed lazily. Annie sang in her kitchen . . . the halter shanks, made impatient, little clinking sounds. *Tang-tang-ting-tang-tang*, went Ab's hammer on the anvil."

It is impossible to realize how diversified were the sounds of the blacksmith without a visit to an active forge. No museum anvil can suggest the sound, for each type of work had its own meters and accents. While on a recording expedition in Europe, we were fortunate in persuading an old Swabian blacksmith and his assistant to fire up their abandoned forge and to demonstrate the techniques. Shaping scythes consisted of a rapid series of taps, followed by slight pauses for inspection. By comparison, the shaping of horseshoes called for the assistant to strike the metal with mighty sledgehammer beats while the smith, with a little hammer for shaping, struck the metal off the beat. The meter was in three, thus:

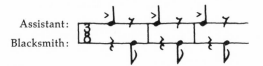

When the smith wanted more flattening he would tap the side of the anvil with two rapid flourishes.

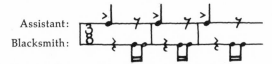

One has to have experienced this to appreciate how deftly the smith would move in and out fashioning the metal between the powerful steady blows of the assistant. We measured the sound at over 100 decibels and residents on the outskirts of the village testified that they used to be able to hear the hammering, which began at dawn and would continue during the harvest season (when scythes had to be flattened regularly) late into the night.

Up to the time of the Industrial Revolution, the sound of the black-smith's hammer was probably the loudest sound a solo human hand ever produced—a brilliant tintinnabulation.

In the Middle East it was the tinsmith's hammer that gave the most strident keynote. They may be heard there still, the happy tinsmiths, squatting in the bazaars, their backs straight as the letter alif, beckoning the visitor with their staccato hammering, which forms a strange counter-point to the phlegmatic shuffling of feet over the uneven stones of the alleys. Today they fashion samovars for tourists; in the past they produced great gongs for the royal armies. In the Orient, the gong served in place of both drum and bell. "We started at break of day from the northern suburb of Ispahan, led by the *chaoûshes* of the pilgrimage, who announced our departure by loud cries and the beating of their copper drums."

Keynote Sounds   Many of the most unique keynote sounds are produced by the materials available in different geographical locales: bamboo, stone, metal or wood, and sources of energy such as water and coal. In older European cities the visitor from abroad is immediately struck by the preponderance of stone. Stone, and objects bumping it, chipping it or scraping it, form the first line of European keynotes. Scott Fitzgerald somewhere comments on "the substantial cobblestones of Zürich," and the way they echoed sound in the narrow streets at night.

In North America wood was a more important keynote, for many towns and cities were carved out of virgin forest. (Wood had, of course, been an original keynote of Europe also; but the forests were depleted when wood was required for the smelting and forging of metals.) The special keynote of British Columbia is still wood. In the early days of Vancouver, wooden planks were used in the construction of sidewalks and streets as well as buildings.

> The first streets were planked and, where necessary, as with the old Water Street, supported on pilings. Photos of the time fail to convey the rumble and roar that quickened the pulse as carriages sped over the timbers. Vancouver had little cobblestone to represent its early paving, and thus the original surface has long since been composted. The sidewalks too were of plank, spaced to the detriment of women's shoe heels.

In those days (1870–1900) some of the Vancouver streets close to the seashore were also paved with clamshells. Wood, especially when elevated on pilings, is a musical surface, for each board gives its own pitch and resonance under boot heel or carriage wheel. Cobblestones possess this quality also, but the drudgery of asphalt and cement is uniform.

Wood meets stone in a combinatory keynote when casks are rolled over cobblestones, a sound which must have created a considerable disturbance in the old days. The city of Cape Town prohibits it (Police Offences Act, No. 27, 1882, para. 27), and it is also prohibited in the Australian city of Adelaide (By-law No. IX, 1934, para. 25a).

A subtle keynote is offered by the sounds of light. Between the soft sniffing of the candle and the stationary hum of electricity a whole chapter in human social history could be written, for the way men light their lives is equally as influential as the way they tell time or write down their languages. (In attributing dynamic social change to the appearance and decline of the printing press, Marshall McLuhan developed only one of several fertile themes.) The potent invention of the mechanical clock is more immediate to our study, but the effect of lighting cannot be ignored.

In the special darkness of the northern winter, where life was centered in small pools of candlelight, beyond which shadows draped and flickered mysteriously, the mind explored the dark side of nature. The underworld creatures of northern mythology are always nocturnal. By candlelight the powers of sight are sharply reduced; the ear is supersensitized and the air stands poised to beat with the subtle vibrations of a strange tale or of ethereal music. . . .

Romanticism begins at twilight—and ends with electricity. By the era of electricity, the last romanticists had folded their wings. Music dismissed the nocturne and the *Nachtstück,* and from the Impressionist salons of 1870 onward, painting emerged into twenty-four-hour daylight.

We will not expect to find striking confessions concerning the sounds of candles or torches among the ancients any more than we find elaborate descriptions among moderns of the 50- or 60-cycle hum; for although both sounds are inescapably there, they are keynotes; and, as I am taking repeated trouble to explain, keynotes are rarely listened to consciously by those who live among them, for they are the ground over which the figure of signals becomes conspicuous. Keynote sounds are, however, noticed when they change, and when they disappear altogether, they may even be remembered with affection. Thus I recall the vivid impression made on me when I first went to Vienna in 1956 and heard the whispering gas lights on the suburban streets; or, on another occasion, the huge hiss of the Coleman lamps in the unelectrified bazaars of the Middle East—which, in the late evening, quite overpowered the bubbling of the waterpipes. Similarly, in a reverse manner, when the heroine in *Doctor Zhivago* first arrived in Moscow after having spent her childhood in the Urals, she was "deafened by the gaudy window displays and glaring lights, as if they too emitted sounds of their own, like the bells and the wheels." In the country, night had been accompanied by "the faint crackling of the wax candles" (Turgenev's phrase), and she was immediately struck by the change. Another example: in his diary of 1919, amidst painterly thoughts, Paul Klee paused to listen when, in his Schwabing apartment, "the asthmatic gas lamp was replaced by a glaring, hissing and spitting carbide lamp."

The Sounds of Night and Day   When towns and cities were dark at night, the sounds of curfew and the voices of night watchmen were important acoustic signals. In London the curfew bell was decreed by William the Conqueror to be rung at eight o'clock. On the first stroke of the signal bell, St. Martin's-le-Grand, all other churches took up the toll and the city gates were closed. Curfew bells were still rung in English towns up to the nineteenth century, as Thomas Hardy records.

> The curfew was still rung in Casterbridge, and it was utilized by the inhabitants as a signal for shutting their shops. No sooner did the deep notes of the bell throb between the housefronts than a clatter of shutters arose through the whole length of the High Street. In a few minutes business at Casterbridge was ended for the day.

In Persian towns curfew was also announced, but the sounds were different.

> I had in succession watched the distant din of the king's band, the crash of the drums, and the swell of the trumpets, announcing sunset.

I had listened to the various tones of the muezzins, announcing the evening prayer; as well as to the small drum of the police, ordering the people to shut their shops, and retire to their homes. The cry of the sentinels on the watch-towers of the king's palace was heard at distant intervals. . . .

After the town settled down for the night, the soundscape, even of a big city like Paris, became hi-fi.

Later that night—last night—when the children and women had quieted down in their back yards enough to let me sleep, I began hearing cabs roll by in the street. They passed only now and then, but after each one I waited for the next in spite of myself, listening for the jingling bell, the clatter of the horse's hooves on the pavement.

Throughout the night, in towns the world over, night watchmen reassured the inhabitants with their punctual sounds.

> Twelve o'clock,
> Look well to your lock,
> Your fire and your light,
> And so good night.

Such was the London cry as recorded by Richard Dering in 1599. Milton records that in his day watchmen carried a bell and chanted a blessing (*Il Penseroso,* line 82 f.). Leigh Hunt has preserved descriptions of several London night watchmen in 1820.

One was a Dandy Watchman, who used to ply at the top of Oxford Street, next the park. We called him the dandy, on account of his utterance. He had a mincing way with it, pronouncing the *a* in the word "past" as it is in *hat,*—making a little preparatory hem before he spoke, and then bringing out his "Păst ten" in a style of genteel indifference, as if, upon the whole, he was of that opinion.

Another was the Metallic Watchman, who paced the same street towards Hanover Square, and had a clang in his voice like a trumpet. He was a voice and nothing else; but any difference is something in a watchman.

A third, who cried the hour in Bedford Square, was remarkable in his calling for being abrupt and loud. There was a fashion among his tribe just come up at that time, of omitting the words "Past" and "o'clock," and crying only the number of the hour.

But by this time the cries of the watchman and the chimes of the town clock were clearly tautological and the watchman was on the wane. Virginia Woolf catches this situation well by placing the watchman sentimentally in the distance. The quote is from *Orlando,* set at about the same period. "There was the faint rattle of a coach on the cobbles. She heard the far-away cry of the night watchman—'Just twelve o'clock on a frosty morning.' No sooner had the words left his lips than the first strike of midnight sounded." Sometimes the watchmen rang bells, sometimes rattles, as reported by Gorky in *The Artamonovs.* Sometimes they blew whistles, and I have heard them blow to one another every fifteen minutes throughout the night in Mexican towns today.

Such nocturnal interruptions were by no means always appreciated; they outraged Tobias Smollett in the eighteenth century.

> . . . I go to bed after midnight, jaded and restless from the dissipations of the day—I start every hour from my sleep, at the horrid noise of the watchmen bawling the hour through every street, and thundering at every door; a set of useless fellows, who serve no other purpose but that of disturbing the repose of the inhabitants.

With the first rays of sunlight the watchman fell silent, and after the introduction of street lighting, he disappeared altogether.

With daybreak a different commotion began. Smollett continues: ". . . and by five o'clock I start out of bed, in consequence of the still more dreadful alarm made by the country carts, and noisy rustics bellowing green pease under my window."

The Keynotes of Horse and Wagon   Smollett was by no means the only commentator to be irritated by the continuous and asymmetrical rattle of brass-bound wheels over cobblestones, and not only Europeans but people living in other parts of the world frequently complained of it. "The creaking of the wheels is indescribable. It is like no sound ever heard in all your life, and makes your blood run cold. To hear a thousand of these wheels all groaning and creaking at one time is a sound never to be forgotten—it is simply hellish." With the carts came the cracking of whips, which the philosopher Arthur Schopenhauer reckoned as the nastiest distraction from intellectual life.

> I denounce it as making a peaceful life impossible; it puts an end to all quiet thought. . . . No one with anything like an idea in his head can avoid a feeling of actual pain at this sudden, sharp crack, which paralyzes the brain, rends the thread of reflection, and murders thought.

That Schopenhauer was not alone in resisting the noise of whips is evident from numerous pieces of legislation, from Europe and abroad, prohibiting "the unnecessary clapping of wagon whips."

One of the most influential keynotes of the early urban soundscape must have been the clatter of horses' hooves, everywhere evident over cobblestone streets, and different from the hollow tramp of hooves on the open ground. Leigh Hunt writes of the night journey by coach when the only way the traveler knew he was passing through a town was the sharpening of the hoofbeat. Returning to the country road, "the moist circuit of the wheels, and the time-beating tread of the horses" eventually hypnotize even the most wakeful passenger.

Surely I am not the only one to conclude that the rhythm of hooves must have knocked around infectiously in the minds of travelers? The influence of horses' hooves on poetical rhythms ought to yield a doctoral thesis or three, and certainly Sir Richard Blackmore once spoke of turning verses "to the rumbling of his coach's wheels." Some equestrian prosodist ought to be able to work the subject up from there. An influence on music is also evident. How else would one care to account for ostinato effects such as the Alberti bass, which came into existence (after 1700) when coach travel throughout Europe became practical, safe and popular? The same influence can be felt in the jigging rhythms of the country square dance, which the southern Americans call, not without reason, "kicker music." Perhaps these thoughts are merely idiosyncratic, but I will stitch them together again when I consider the influence of the railroad on jazz and the automobile on contemporary music.

The Rhythms of Work Begin to Change   Before the Industrial Revolution work was often wedded to song, for the rhythms of labor were synchronized with the human breath cycle, or arose out of the habits of hands and feet. We will later discuss how singing ceased when the rhythms of men and machines got out of sync, but it is not too early to point out the tragedy. Before this, the shanties of the sailor, the songs of the fields and workshops set the rhythm, which street vendors and flower girls imitated or counterpointed in a vast choral symphony. At first, as Gorky's novel *The Artamonovs* testifies, the workers brought their songs to the cities willingly enough.

> Pyotr Artamonov paced the building site, pulling absently at his ear, observing the work. A saw ate lusciously into wood; planes shuffled, wheezing, to and fro; axes tapped loud and clear; mortar splashed wetly onto masonry, and a whetstone sobbed against a dull axe edge. Carpenters, lifting a beam, struck up *Dubinushka,* and somewhere a young voice sang out lustily:
>
> "Friend Zachary visited Mary,
> Punched her mug to make her merry."

Later, the workers would resign themselves spitefully to jeering at the mill. Then they would just

> gather on the bank of the Vataraksha, nibbling at pumpkin and sunflower seeds and listening to the snorting and whining of the saws, the shuffling of the planes, the resounding blows of the sharp axes. They

would speak in mocking tones of the fruitless building of the tower of Babel.

The industrial workshop killed singing. As Lewis Mumford put it in *Technics and Civilization:* "Labor was orchestrated by the number of revolutions per minute, rather than the rhythm of song or chant or tattoo."

Street Criers   But this came later. Before the Industrial Revolution the streets and workshops were full of voices, and the farther south one went in Europe the more boisterous they appeared to become.

> Turn your eyes upward, myriads of windows and balconies, curtains swinging in the sun, and leaves and flowers and among them, people, just to confirm your illusion. Cries, screams, whipcracks deafen you, the light blinds you, your brain begins to feel dizzy and you gulp air. You feel drawn into becoming part of the enthusiastic demonstration, to applaud, to cry "Evvive"—but for what? What is there before your eyes is nothing exceptional or extraordinary. All is perfectly calm; no deep political passion is stirring in these people. They all mind their business and talk about normal things; it is just a day like any other. It is Naples' life in its perfect normality, nothing more.

Why do the voices of South Europeans always seem louder than their northern neighbors? Is it because they spend more time outdoors where the ambient noise level is higher? We recall that the Berbers learned to shout because they had to shout over the cataracts of the Nile.

But the streets of all major European towns were seldom quiet in those days, for there were the constant voices of hawkers, street musicians and beggars. The beggars in particular plagued the composer Johann Friedrich Reichardt when he visited Paris in 1802–3. "Usually they are not violent as they fall on one, but they hamper one and touch the heart all the more with their continuous beseeching cries and their miserable behavior." The ubiquitous street cries were impossible to avoid. "The uproar of the street sounded violently and hideously cacophonous," reported Virginia Woolf in *Orlando;* but this is too general. Actually each hawker had an uncounterfeiting cry. More than the words, it was the musical motif and the inflection of the voice, passed in the trade from father to son, that gave the cue, blocks away, to the profession of the singer. In the days when shops moved on wheels, ads were vocal displays. Street cries attracted the attention of composers and were incorporated into numerous vocal compositions, by Janequin in sixteenth-century France and by Weelkes, Gibbons and Dering in the England of Shakespeare's time. The Fancies by the last three composers contain some one hundred fifty different cries and itinerant vendors' songs. A catalogue of some of these gives a good idea of the variety of goods and services which were available in the towns of Elizabethan England:

13 different kinds of fish,
18 different kinds of fruit,
 6 kinds of liquors and herbs,
11 vegetables,
14 kinds of food,
14 kinds of household stuff,
13 articles of clothing,
 9 tradesmen's cries,
19 tradesmen's songs,
 4 begging songs for prisoners,
 5 watchman's songs,
 1 town crier.

The town crier preserved by Dering is clearly from before the days of the Puritan reforms. He begins with the traditional invocation "Oyez," from the Norman French verb *oüir,* to hear.

> Oyez, Oyez. If any man or woman, city or country, that can tell any tidings of a grey mare with a black tail, having three legs and both her eyes out, with a great hole in her arse, and there your snout, if there be any that can tell any tidings of this mare, let him bring word to the crier and he shall be well pleas'd for his labour.

The practice of maintaining town criers proceeded down to about 1880, or at least that is the time when their names disappear from the directories of cities like Leicester.

Public hawking was carried on also in the theaters and opera houses, as Johann Friedrich Reichardt reported from Paris.

> Between the acts, hawkers enter bearing orangeade, lemonade, ice cream, fruit, and so forth, while others bring opera libretti, programs, evening newspapers and journals and still others advertise binoculars, all vying with one another and making such a commotion that one is driven to distraction. This is even worse on those days when the theatre, as is common in France, is so full that the musicians of the orchestra are forced out to accommodate extra spectators. Right after the last word of the tragedy, the hawkers push past the doors and bawl out "Orangeade, Lemonade, Glacés! marchand des lorgnettes!" and so forth, completely bereft of any music, lacerating the ears and feelings of all sensitive spectators.

Noise in the City

It will be noticed, from several of the quotations of the last few pages, that street music was a continual subject of controversy. Intellectuals were irritated by it. Serious musicians were outraged—for frequently it appears that unmusical persons would engage in the practice, not at all to bring pleasure, but merely to have their silence bought off. But resistance moved to the middle class as well, as soon as it contemplated an elevation of life style. After art music moved indoors, street music became an object of increasing scorn, and a study of European noise abatement legislation between the sixteenth and nineteenth centuries shows how increasing amounts of it were directed against this activity. In England, street music was suppressed by two Acts of Parliament during the reign of Elizabeth I, but it can hardly have been very effective. Hogarth's well-known eighteenth-century print, *The Enraged Musician,* shows the conflict between indoor and outdoor music in full view. By the nineteenth century, a by-law in Weimar had forbidden the making of music unless conducted behind closed doors. The bourgeoisie was gaining the upper hand, on paper at least. In England the brewer and Member of Parliament, Michael T. Bass, published a book in 1864 entitled *Street Music in the Metropolis,* together with a proposed Bill, designed to put an end to the abuse. Bass received a great many letters and petitions supporting his Bill, including one signed by two hundred "leading composers and professors of music of the metropolis," who complained vigorously of the way in which "our professional duties are seriously interrupted." Another letter, signed by Dickens, Carlyle, Tennyson, Wilkie Collins, and the Pre-Raphaelite painters John Everett Millais and Holman Hunt, stated:

> Your correspondents are, all, professors and practitioners of one or other of the arts or sciences. In their devotion to their pursuits— tending to the peace and comfort of mankind—they are daily interrupted, harassed, worried, wearied, driven nearly mad, by street musicians. They are even made especial objects of persecution by brazen performers on brazen instruments, beaters of drums, grinders of organs, bangers of banjos, clashers of cymbals, worriers of fiddles, and bellowers of ballads; for, no sooner does it become known to those producers of horrible sounds that any of your correspondents have particular need of quiet in their own houses, than the said houses are beleaguered by discordant hosts seeking to be bought off.

A further communication received by Bass for his proposed bill was in the form of a detailed list of interruptions from Charles Babbage, the eminent mathematician and inventor of the calculating machine. Brass bands, organs and monkeys were the chief distractions, and Babbage came to the conclusion that "one-fourth part of my working power has been destroyed by the nuisance against which I have protested."

Selective Noise Abatement: The Street Crier Must Go

As a result of this agitation, the Metropolitan Police Act of 1864 was passed, though the problem cannot have been immediately solved, for street cries continued to be noted until the turn of the century and later. But by 1960, the only European city in which street cries could still regularly be heard was Istanbul. When at last the legislators of European towns were able to conclude that the problem of street music had been solved, they failed to appreciate the correct reason for it. It was not the result of centuries of legislative refinement but the invention of the automobile that muffled the voices of the street cries. Then slow-witted administrations all over the world got down to designing by-laws to solve a problem that had already disappeared. "No hawker, huckster, peddler or petty chapman, news vendor or other person shall by his intermittent or reiterated cries disturb the peace, order, quiet or comfort of the public" (Vancouver, By-Law No. 2531, passed in 1938).

By the 1930s Parisian citizens were lamenting the disappearance of street criers—*si la chanson française ne doit pas mourir ce sont les chanteurs des rues qui doivent la perpétuer;* but Professor Beauty was by that time in his padded cell, which is to say that the disappearance of street music has been largely a matter of indifference to aesthetes and collectors.

The study of noise legislation is interesting, not because anything is ever really accomplished by it, rather because it provides us with a concrete register of acoustic phobias and nuisances. Changes in legislation give us clues to changing social attitudes and perceptions, and these are important for the accurate treatment of sound symbolism.

Early noise abatement legislation was selective and qualitative, contrasting with that of the modern era, which has begun to fix quantitative limits in decibels for all sounds. While most of the legislation of the past was directed against the human voice (or rather the rougher voices of the lower classes), no piece of European legislation was ever directed against the far larger sound—if objectively measured—of the church bell, nor against the equally loud machine which filled the church's inner vaults with music, sustaining the institution imperiously as the hub of community life—until its eventual displacement by the industrialized factory.

PART TWO

The Post-Industrial
Soundscape

FIVE

~~~~~~

The Industrial Revolution

The Lo-Fi Soundscape of the Industrial Revolution The lo-fi soundscape was introduced by the Industrial Revolution and was extended by the Electric Revolution which followed it. The lo-fi sound-scape originates with sound congestion. The Industrial Revolution intro-duced a multitude of new sounds with unhappy consequences for many of the natural and human sounds which they tended to obscure; and this development was extended into a second phase when the Electric Revolu-tion added new effects of its own and introduced devices for packaging sounds and transmitting them schizophonically across time and space to live amplified or multiplied existences.

Today the world suffers from an overpopulation of sounds; there is so much acoustic information that little of it can emerge with clarity. In the ultimate lo-fi soundscape the signal-to-noise ratio is one-to-one and it is no longer possible to know what, if anything, is to be listened to. This, in summary, is the transformation of the soundscape which we will study in the next chapters.

The Industrial Revolution in England, the country which, for a variety of reasons, became the first to mechanize, took place approximately be-tween 1760 and 1840. The principal technological changes which affected the soundscape included the use of new metals such as cast iron and steel as well as new energy sources such as coal and steam.

The textile industry was the first to undergo industrialization. John Kay's flying shuttle (1733), James Hargreaves's spinning jenny (1764–69) and Richard Arkwright's waterframe (1769) led to the development of the power loom by 1785. Increased production of finished cotton goods led to a greater demand for raw cotton, a problem which was solved in the U.S.A. by Eli Whitney's cotton gin (1793). Other industries quickly followed, for as Alfred North Whitehead observed: "The greatest invention of the nine-teenth century was the invention of the method of invention." A list of some of the more outstanding eighteenth-century inventions will allow the imaginative reader to overhear the changes in the soundscape which were worked by the new materials under the impress of new energy sources and the relentless precision of new machinery.

1711:	Sewing machine
1714:	Typewriter
1738:	Cast-iron rail tramway (at Whitehaven, England)
1740:	Cast steel
1755:	Iron wheels for coal cars
1756:	Cement manufacture
1761:	Air cylinders; piston worked by water wheel; more than tripled production of blast furnace
1765–69:	Improved steam pumping engine with separate condenser
1767:	Cast-iron rails (at Coalbrookdale)
1774:	Boring machine
1775:	Reciprocative engine with wheel
1776:	Reverberatory furnace
1781–86:	Steam engine as prime mover
1781:	Steamboat
1785:	First steam spinning mill (at Papplewick)
1785:	Power loom
1785:	Screw propeller
1787:	Iron steamship
1788:	Threshing machine
1790:	Sewing machine first patented
1791:	Gas engine
1793:	Signal telegraph
1795–1809:	Food canning
1796:	Hydraulic press
1797:	Screw-cutting lathe

The social concomitants to these changes were also profound. Agricul-tural workers were disfranchised and sent to the cities to seek work in the factories. Operated by steam engines, lighted by gas, the new factories could work nonstop day and night, and pauperized workers were forced to do the same. The working day was increased to sixteen hours or more, with a single hour off for dinner. Workers lived in squalid quarters near the factories, cut off from the countryside, with almost no recreational facilities except the public houses; and these, if we accept the evidence of numerous earwitnesses, became centers of much greater noise and rowdi-ness during the eighteenth century than before.

I have already mentioned how factories put an end to the unity of work and song. At a later date, after the reform work of men like Robert Owen, the urge for singing reappeared in the British choral societies, which flourished best in the factory towns of the North. Workmen who experienced the crucifixion of human culture then sang *Messiah* at Christmas in thousand-voice choirs.

The cacophonies of iron pushed out over the countryside first in the form of the railroad and the threshing machine. We can measure the phases of change as the new farming machinery moved out from England across Europe. While Tolstoy's Russian peasants still continued to sing over their sickles, the heroine of Hardy's *Tess of the d'Urbervilles* (contemporary of Anna Karenina) stands mutely over her work smothered by the concatenated roar of the threshing machine.

> A hasty lunch was eaten as they stood, without leaving their positions, and then another couple of hours brought them near to dinnertime; the inexorable wheels continuing to spin, and the penetrating hum of the thresher to thrill to the very marrow all who were near the revolving wirecage.

The Sounds of Technology
Sweep Across Town and Country
While the philosophy of utilitarianism was sufficient to condone the inhumanities of Coketown, the machine was immediately conspicuous when it was introduced into provincial life. It took time for the sounds of technology to rub their way across Europe. The following set of earwitness accounts by writers over several generations reveals how the new sounds were gradually accepted as inevitable.

French towns were upset at first by the new rhythms and aberrational noises of the machine, as Stendhal makes clear on one of the first pages of *The Red and the Black* (1830).

> The little town of Verrières must be one of the prettiest in the Franche-Comté. Its white houses with their steep, red tile roofs spread across a hillside, the folds of which are outlined by clumps of thrifty chestnut trees. The Doubs flows a couple of hundred feet below the town's fortifications, built long ago by the Spaniards and now fallen into ruins....
>
> Scarcely inside the town, one is stunned by the racket of a roaring machine, frightful in its appearance. Twenty ponderous hammers, falling with a crash which makes the street shudder, are lifted for each new stroke by the power of a water wheel. Every one of these hammers makes, every day, I don't know how many thousand nails. Young, pretty, fresh-faced girls, slip little slivers of iron into place beneath the sledge hammers, which promptly transform them into nails.

By 1864 French towns were alive with factories, and were described with disdain by the Goncourts.

> A vague, indeterminate smell of grease and sugar, mixed with the emanations from the water and the smell of tar, rose from the candle factories, the glue factories, the tanneries, the sugar refineries, which were scattered about on the quay amongst thin, dried-up grass. The noise of foundries and the screams of steamwhistles broke, at every moment, the silence of the river.

By the early twentieth century the sounds of technology became more acceptable to the urban ear, "blending" with the natural rhythms of antiquity. As Thomas Mann described it:

> We are encompassed with a roaring like that of the sea; for we live almost directly on the swift-flowing river that foams over shallow ledges at no great distance from the poplar avenue.... There is a locomotive foundry a little way downstream. Its premises have been lately enlarged to meet increased demands, and light streams all night long from its lofty windows. Beautiful glittering new engines roll to and fro on trial runs; a steam whistle emits wailing head-tones from time to time; muffled thunderings of unspecified origin shatter the air.... Thus in our half-suburban, half-rural seclusion the voice of nature mingles with that of man, and over all lies the bright-eyed freshness of the new day.

Ultimately the throb of the machine began to intoxicate man everywhere with its incessant vibrations. D. H. Lawrence (1915): "As they worked in the fields, from beyond the now familiar embankment came the rhythmic run of the winding engines, startling at first, but afterwards a narcotic to the brain."

Eventually then the noises of modern industrial life swung the balance against those of nature, a fact which the futurist, Luigi Russolo, was the first to point out in his manifesto *The Art of Noises* (1913). Writing on the eve of the First World War, Russolo excitedly proclaimed that the new sensibility of man depended on his appetite for noises, which would achieve their grandest opportunity for expression in mechanized warfare.

Noise Equals Power   We have gone far enough to show how the soundscape of both city and country was being transmogrified during the eighteenth and nineteenth centuries. We are now confronted by an enigma: despite the vast increase in noise that the new machines created, rarely do we find opposition to these noises.

In England, the first criticism of working conditions in factories was that of Sadler's Factory Investigating Committee of 1832. This pathetic seven-hundred-page document is filled with hideous descriptions of brutality and human degradation—shifts extending to thirty-five hours, children sleeping in the mills in order not to be late for work, workers collapsing into the machines from sheer fatigue, alcoholism among children—but nowhere is noise mentioned as a factor contributing to the tragedy of these environments. Only once or twice does one encounter there a reference to the "rumbling noise" of the machinery. When sound is noticed it is usually the screams of the workers when they are beaten.

> I happened to be at the other end of the room, talking; and I heard the blows, and I looked that way, and saw the spinner beating one of the girls severely with a large stick. Hearing the sound led me to look round, and to ask what was the matter, and they said it was "Nothing but —— paying [beating] his ligger-on."

The only time the machines were ever stopped was to impress visitors, or during meal breaks, when the children had to clean them on their own time. Otherwise they rattled on undetected, and Sadler's interviewees even spoke of the "silence" of the mills, by which they meant the "rule of silence." "Is one part of the discipline of these mills profound silence?—Yes, they will not allow them to speak; if they chance to see two speaking, they are beaten with the strap."

The only people to criticize the "prodigious noise" of machinery were the writers, figures like Dickens and Zola. Dickens, in *Hard Times* (1854):

> Stephen bent over his loom, quiet, watchful, and steady. A special contrast, as every man was in the forest of looms where Stephen worked, to the crashing, smashing, tearing piece of mechanism at which he laboured.

Zola, in *Germinal* (1885):

> And now it had occurred to him to open the steam-cocks and let out the steam. The jets went off like gunshots and the five boilers blew off like hurricanes, with such a thunderous hissing that your ears seemed to be bleeding.

Despite these attacks, it was still to be a hundred years before noise criteria would be established and enforced as part of hygiene programs in industry. Neither unions nor social reformers nor the medical profession caught the theme. Noise was certainly known to cause deafness as early as 1831 when Fosbroke described deafness occurring among blacksmiths, but this remained an isolated study until 1890 when Barr surveyed one hundred boilermakers and discovered that not one of them had normal hearing.* Hammering and riveting steel plates together produced an intense noise, resulting in a form of hearing impairment in which there is

---

*The earliest study of industrial deafness that I have been able to discover was that of Bernardino Ramazzini, *Diseases of Workers* (*De Morbis Artificium*), 1713.

deafness to high frequencies. The term "boilermaker's disease" came into use shortly afterward to refer to all kinds of industrial hearing loss, though its prevention only received serious consideration in most industrialized countries toward 1970.

The inability to recognize noise during the early phases of the Industrial Revolution as a factor contributing to the multiplicatory toxicity of the new working environments is one of the strangest facts in the history of aural perception. We must try to determine the reason. It may be partly explained as a result of the inability to measure sounds quantitatively. A sound might be recognized as unpleasantly loud, but until Lord Rayleigh built the first practical precision instrument for the measurement of acoustic intensity in 1882, there was no way of knowing for certain whether a subjective impression had an objective basis. The decibel, as a means of establishing definite sound pressure levels, did not come into extended use until 1928.

But I want to extend a thought which I had begun to develop in Part One. We have already noted how loud noises evoked fear and respect back to earliest times, and how they seemed to be the expression of divine power. We have also observed how this power was transferred from natural sounds (thunder, volcano, storm) to those of the church bell and pipe organ. I called this Sacred Noise to distinguish it from the other sort of noise (with a small letter), implying nuisance and requiring noise abatement legislation. This was always primarily the rowdy human voice. During the Industrial Revolution, Sacred Noise sprang across to the profane world. Now the industrialists held power and they were granted dispensation to make Noise by means of the steam engine and the blast furnace, just as previously the monks had been free to make Noise on the church bell or J. S. Bach to open out his preludes on the full organ.

The association of Noise and power has never really been broken in the human imagination. It descends from God, to the priest, to the industrialist, and more recently to the broadcaster and the aviator. The important thing to realize is this: to have the Sacred Noise is not merely to make the biggest noise; rather it is a matter of having the authority to make it without censure.

Wherever Noise is granted immunity from human intervention, there will be found a seat of power. The noisy clank of Watt's original engine was maintained as a sign of power and efficiency, against his own desire to eliminate it, thus enabling the railroads to establish themselves more emphatically as the "conquerers" that I will, in a moment, let Charles Dickens describe. A glance at the sound output of any representative selection of modern machines is enough to indicate where the centers of power lie in the modern world.

Steam engine	85 dBA
Printing works	87 dBA
Diesel-electric generator house	96 dBA

Screw-heading machine	101 dBA
Weaving shed	104 dBA
Sawmill chipper	105 dBA
Metalwork grinder	106 dBA
Wood-planing machine	108 dBA
Metal saw	110 dBA
Rock band	115 dBA
Boiler works, hammering	118 dBA
Jet taking off	120 dBA
Rocket launching	160 dBA

Sound Imperialism The historian Oswald Spengler distinguishes two phases in the development of a social movement: the cultural phase, during which the main ideas are still maturing; and the civilization phase, during which the main ideas, having matured, are legalized and transmitted abroad. Imperialism is the word used to refer to the extension of an empire or ideology to parts of the world remote from the source. It is Europe and North America which have, in recent centuries, masterminded various schemes designed to dominate other peoples and value systems, and subjugation by Noise has played no small part in these schemes. Expansion took place first on land and sea (train, tank, battleship) and then in the air (planes, rocketry, radio). The moon probes are the most recent expression of the same heroic confidence that made Western Man a world colonial power.

When sound power is sufficient to create a large acoustic profile, we may speak of it, too, as imperialistic. For instance, a man with a loudspeaker is more imperialistic than one without because he can dominate more acoustic space. A man with a shovel is not imperialistic, but a man with a jackhammer is because he has the power to interrupt and dominate other acoustic activities in the vicinity. (In this sense we note that outside workers were able to improve their position remarkably after they were in possession of tools to attract attention to themselves. No one listens to a ditchdigger.) Similarly, the growing importance of the international aviation industry can be easily assessed from the growth patterns of airport noise profiles. Western Man leaves his calling cards all over the world in the form of Western-made or Western-inspired machinery. As the factories and the airports of the world multiply, local culture is pulverized into the background. Everywhere one travels today one hears the evidence, though only in the more remote places is the incongruity immediately striking.

Increase in the intensity of sound output is the most striking characteristic of the industrialized soundscape. Industry must grow; therefore its sounds must grow with it. That is the fixed theme of the past two hundred years. In fact, noise is so important as an attention-getter that if quiet machinery could have been developed, the success of industrialization might not have been so total. For emphasis let us put this more dramatically: if cannons had been silent, they would never have been used in warfare.

The Flat Line in Sound   The Industrial Revolution introduced another effect into the soundscape: the flat line. When sounds are projected visually on a graphic level recorder, they may be analyzed in terms of what is called their envelope or signature. The principal characteristics of a sound envelope are the attack, the body, the transients (or internal changes) and the decay. When the body of the sound is prolonged and unchanging, it is reproduced by the graphic level recorder as an extended horizontal line.

Machines share this important feature, for they create low-information, high-redundancy sounds. They may be continuous drones (as in a generator); they may be rough-edged, possessing what Pierre Schaeffer calls a "grain" (as in mechanical sawing or filing); or they may be punctuated with rhythmic concatenations (as in weaving or threshing machines) —but in all cases it is the continuousness of the sound which is its predominating feature.

The flat continuous line in sound is an artificial construction. Like the flat line in space, it is rarely found in nature. (The continuous stridulation of certain insects like cicadas is an exception.) Just as the Industrial Revolution's sewing machine gave us back the long line in clothes, so the factories, which operated night and day nonstop, created the long line in sound. As roads and railroads and flat-surfaced buildings proliferated in space, so did their acoustic counterparts in time; and eventually flat lines in sound slipped out across the countryside also, as the whine of the transport truck and the airplane drone demonstrate.

A few years ago, while listening to the stonemasons' hammers on the Takht-e-Jamshid in Teheran, I suddenly realized that in all earlier societies the majority of sounds were discrete and interrupted, while today a large portion—perhaps the majority—are continuous. This new sound phenomenon, introduced by the Industrial Revolution and greatly extended by the Electric Revolution, today subjects us to permanent keynotes and swaths of broad-band noise, possessing little personality or sense of progression.

Just as there is no perspective in the lo-fi soundscape (everything is present at once), similarly there is no sense of duration with the flat line in sound. It is suprabiological. We may speak of natural sounds as having biological existences. They are born, they flourish and they die. But the generator or the air-conditioner do not die; they receive transplants and live forever.

The flat line in sound emerges as a result of an increased desire for speed. Rhythmic impulse plus speed equals pitch. Whenever impulses are speeded up beyond 20 occurrences or cycles per second, they are fused together and are perceived as a continuous contour. Increased efficiency in manufacturing, transportation and communication systems fused the impulses of older sounds into new sound energies of flat-line pitched noise. Man's foot sped up to produce the automobile drone; horses' hooves sped up to produce the railway and aircraft whine; the quill pen sped up to produce the radio carrier wave, and the abacus sped up to produce the whirr of computer peripherals.

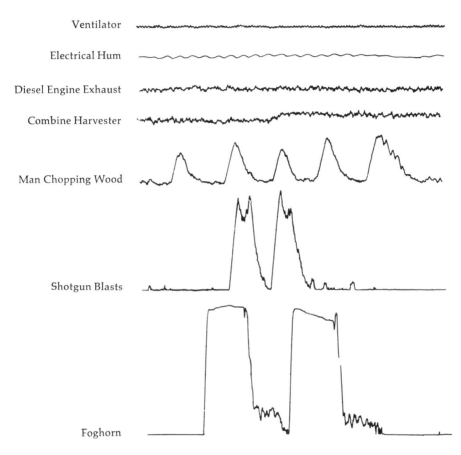

| Ventilator |
| Electrical Hum |
| Diesel Engine Exhaust |
| Combine Harvester |
| Man Chopping Wood |
| Shotgun Blasts |
| Foghorn |

Graphic level recordings of typical flat-line and impact sounds.

Henri Bergson once asked how we should know about it if some agent suddenly doubled the speed of *all* events in the universe? Quite simply, he replied, we should discern a great loss in the richness of experience. Even as Bergson wrote, this was happening, for as discrete sounds gave way to flat lines, the noise of the machine became "a narcotic to the brain," and listlessness increased in modern life.

The function of the drone has long been known in music. It is an anti-intellectual narcotic. It is also a point of focus for meditation, particularly in the East. Man listens differently in the presence of drones, and the importance of this change in perception is becoming evident in the West.

The flat line in sound produces only one embellishment: the glissando
—that is, as the revolutions increase the pitch gradually rises, and as they
decrease the pitch descends. Then flat lines become curved lines. But they
are still without sudden surprises. When flat lines become jerky or dotted
or looping lines—the machinery is falling apart.

Another type of curve produced by the flat line is the Doppler effect,
which results when a sound is in motion at sufficient velocity to cause a
bunching up of the sound waves as the sound approaches an observer
(resulting in a rise in pitch) and an elongation of the sound waves as the
sound recedes (resulting in a lowering of pitch). There are certainly Doppler effects in nature (the flight of a bee, for instance, or the galloping of
horses) but only after the new speeds of the Industrial Revolution did the
effect become conspicuous enough to be "discovered." Christian Johann
Doppler (1803–53) formulated the explanation of the effect to which he
has bequeathed his name in a work entitled *Über das Farbige Licht der
Doppelsterne,* where he applied the principle to light waves. But Doppler
acknowledged that he worked by analogy from sound to light.

Some sounds move in space, some do not; and we may move some
sounds by carrying them with us. But which sound attracted Doppler's
ear? It could only have been the railway. Although he does not mention
this, we do know trains were used to verify the Doppler effect. About 1845
"musically trained observers were stationed along the tracks of the Rhine
Railroad between Utrecht and Maarsen in Holland and listened to trumpets played in a railway car as it sped past. From the known pitch of the
trumpet and the apparent pitch of the approaching and receding tones, the
speed of the railway car was estimated with fairly good accuracy."

The Lore of Trains   The first railway was the Stockton and Dar-
lington run in England (1825), designed to carry coal from the mines to the
waterways. It proved so immediately successful that within a few years
Britain was covered with a railway network. Dickens described the new
sound in 1848:

> Night and day the conquering engines rumbled at their distant work,
> or, advancing smoothly to their journey's end, and gliding like tame
> dragons into the allotted corners grooved out to the inch for their
> reception, stood bubbling and trembling there, making the walls
> quake, as if they were dilating with the secret knowledge of great
> powers yet unsuspected in them, and strong purposes not yet
> achieved.

From England the railway system fanned out quickly across Europe
and the world. France had a railway by 1828 as did the U.S.A., Ireland by
1834, Germany by 1835, Canada by 1836, Russia by 1837, Italy by 1839,
Spain by 1848, Norway and Australia by 1854, Sweden by 1856 and Japan
by 1872.

The train conquered the world with a minimum of opposition. Dickens didn't like it: "Louder and louder yet, it shrieks and cries as it comes tearing on resistless to the goal." Neither did Wagner, and although the Bavarian College of Medicine protested in 1838 that the speed with which trains traveled would undoubtedly cause brain damage, the trains remained and the tracks multiplied.

Of all the sounds of the Industrial Revolution, those of trains seem across time to have taken on the most attractive sentimental associations. J. M. W. Turner's celebrated painting *Rain, Steam and Speed* (1844), with its locomotive thrusting down diagonally on the spectator, was the first lyric inspired by the steam engine. It was a painter, too, who caught the next change in the epic of the railroads. By 1920 the main lines of Europe (though not of England and North America) were being electrified, and the change is recorded in de Chirico's wistful landscapes, where silent smoke-puffing trains pass out of sight in the extreme distance.

By comparison with the sounds of modern transportation, those of the trains were rich and characteristic: the whistle, the bell, the slow chuffing of the engine at the start, accelerating suddenly as the wheels slipped, then slowing again, the sudden explosions of escaping steam, the squeaking of the wheels, the rattling of the coaches, the clatter of the tracks, the thwack against the window as another train passed in the opposite direction— these were all memorable noises.

The sounds of travel have deep mysteries. Just as the post horn had once carried the imagination over the horizon, so also did its replacement, the train whistle. On European trains the whistle is high and piping: "Then the shrill whistle of the trains reechoed through the heart, with fearsome pleasure, announcing the far-off come near and imminent."

In North America, on the other hand, the whistle is low and powerful, the utterance of a big engine with a heavy load. On the prairies—so flat that one can see the full train from engine to caboose, spread out like a stick across the horizon—the periodic whistlings resound like low, haunting moans. "The Canadian train whistle sounds like a dejected monster. It wails, and the pitch descends, unlike that of our British trains which rises in a chirpy and optimistic manner. The Canadian whistle sounds as if it has travelled far and still has a long way to go." Farmers knew how to interpret these sounds. "When the train's whistle sounds hollow, the weather will turn colder," runs an Ontario proverb. The train's whistle was the most important sound in the frontier town, the solo announcement of contact with the outside world. It was the stop clock of the elementary community, as predictable and reassuring as the church bell. In those days trains spoke to the heart of every man, and small boys came to greet the panting engine.

Trains speak to each other too. Each railroad employs a binary code of whistle signals by which quite precise messages can be communicated. But unlike the signals of the post horn, which we are given to understand everyone knew, the language of the trains is a mystery code, known to the

trainmen alone. Even without knowing the codes, those who listen atten-
tively to the soundscape will notice the personality and style each engineer
manages to bring to these elementary utterances. Some slur the signals,
barely distinguishing the articulations; others separate each blast with
lengthy pauses. With considerable artistry others manage to get the notes
to slide in pitch by careful manipulation of the control valve. This last style
of performance is atavistic, carrying us back to the old steam whistle which
was naturally tapered at the edges. The original steam whistle was three-
toned. Part of the fame of the legendary American engineer Casey Jones
was acoustic, for Casey had a special five-tone whistle which he carried
with him wherever he went.

Besides the variations in rhythm and articulation, the listener will also
notice differences in sound quality and pitch. While the old steam whistles
produced a cluster of frequencies, many modern whistles, especially on
diesel engines, are single tones. Others are diads or triads, tuned in the
factory, sometimes to the specifications of the customer. While American
railroads have preferred the single pitch, Canadian railroads have with-
drawn this type of whistle due to the number of level-crossing accidents
attributed to it during the changeover period from steam to diesel engines.
An attempt to reproduce the quality of the original steam whistle resulted
in the adoption of specifically tuned air horns, one version of which, now
used by the Canadian Pacific Railroad, is the E-flat minor triad in root
position with the tonic pitched at 311 hertz. This deep and haunting
whistle, sounded by every train, who knows how many thousands of times
during the long haul from Atlantic to Pacific through spectacular and
lonely countryside, provides the unifying soundmark of the nation. More
than any other sound it is uncounterfeitingly Canadian.

> The improvement in city conditions by the general
> adoption of the motorcar can hardly be overestimated.
> Streets—clean, dustless and odorless—with light rub-
> ber-tired vehicles moving swiftly and *noiselessly* over
> their smooth expanse would eliminate a greater part of
> the nervousness, distraction and strain of modern life.
> *Scientific American,* July, 1899

The Internal Combustion Engine   The internal combustion
engine now provides the fundamental sound of contemporary civilization.
It is the keynote, as surely as water was the keynote of thalassocratic
civilization, and the wind is the keynote of the steppes.

In the external combustion engine a load of water is mixed with a load
of coal to produce driving energy. Coal and water are bulky and heavy. The
steam locomotive was accordingly confined to public enterprises. The internal combustion engine is light and easy to operate; it transferred power
to the individual. In industrially advanced societies the average citizen may
operate several internal combustion engines in the course of an average day

(car, motorcycle, truck, power lawnmower, tractor, generator, power tools, etc.) and the sound will be heard in his ear many hours each day.

By 1970, the United States was producing more automobiles annually than babies, but the Asian market still looked encouraging. An advertisement in *The New Yorker* magazine that year showed the globe with every available land mass covered with Hertz rent-a-cars. By that date classic cities of gems and germs like Istanbul and Isfahan had also become cities of incredible traffic jams. The reason for this was not merely the volume of traffic, but the way in which vehicles were driven. In order for a society to obey traffic codes it must have survived two important experiences: the Industrial Revolution and mechanized warfare. Americans can drive on the "belt" road (note the name) around Washington with great adroitness, but the Asian still drives his car as if it were a camel or a mule. Stoplights are ignored and the horn becomes a whip with which to cajole and punish the stubborn animal.

When two swaths of broad-band noise of the same intensity are superposed the result is an increase of approximately three decibels. Two cars, each producing a sound of 80 decibels, thus give a sound of 83 decibels. Assuming constant engine noise, each doubling of production in the automobile industry would give an elevation of three decibels of broad-band noise to the sonic environment. In fact, automobile engines are not uniformly constructed. The American manufacturers, for instance, produced their quietest automobiles around 1960. During the sixties they began to get louder again. By 1971 the Detroit manufacturers had begun to make the increased noise of their machines an advertised feature. The following is a magazine advertisement:

THE
1971
MUSCLE
CARS
This sleek, high-powered monster is
American Motors' 7 Javelin AMX.
Press the accelerator,
it roars.

That year General Motors informed us that

> ... the trend toward large displacement engines and higher compression ratios makes for increased engine noise, induction, and exhaust noise ... higher compression ratios ... result in larger deformations of the engine-block structure and, hence, higher radiated noise levels. ... We feel, on the basis of many cars, that muffler design and performance have nicely kept pace with requirements.

Today the value of the automobile is under serious scrutiny. As local noise abatement by-laws and practices seek to reduce its sound output by setting increasingly tough noise standards, in the end perhaps only energy shortages will silence it. As the automobile becomes obsolete, its rattle becomes deathening.

Sheer volume aside, the human sound which most closely approximates that of the internal combustion engine is the fart. The analogies between the automobile and the anus are conspicuous. First of all the exhaust pipe is placed at the rear, at the same position as the rectum in animals. Cars are also stored in dirty and dark underground garages, beneath the haunches of the modern dwelling. Freud says there are anal types. There are probably also anal eras.

The Growth of Muscle Sounds

Someone once observed—I think it was Aldous Huxley—that for contemporary urban man half the imagery of traditional poetry was lost. The same thing is happening to the soundscape, where the sounds of nature are being lost under the combined jamming of industrial and domestic machinery. More is less. A couple of illustrations from close to home will suffice to illustrate the equation.

In 1959 Canada manufactured $8,596,000 worth of power chainsaws; by 1969 this had risen to $26,860,000. The power chainsaw produces a sound of between 100 and 120 dBA, giving it a sweepout in a quiet forest of 8 to 10 square kilometers. It is possible to theorize that by 1974 the combined ripping of the 316,781 power chainsaws produced that year alone, if operated simultaneously, could cover about one-third of Canada's 9,222,977 square kilometers with their sound.

A West Coast Indian girl taught me how to listen to the voices of the trees through the bark of their trunks. "They tell the story of my people," she said. When the white men arrived in British Columbia, they could not teach the Indians to use the mechanical saw, or to fell trees in such a way that one tree could be made to knock down four others—the so-called domino technique. When the spirit of the deity inhabits the tree, one hesitates. Today, as the jabberware of the forest industry bevels down the woods, no one hears the frightened cries of the tree victims.

"If a tree might move by foot or wing it would not suffer the pain of the saw or the blows of the axe," wrote Rumi in the thirteenth century. As a matter of fact, we do know that trees and other vegetables tremble and send out electrical emergency charges before they are cut.

The snowmobile will serve as our other example of the devastating effect the careless introduction of technology can have on a society. The snowmobile, a Canadian discovery, is a recent invention but its ramming has already transmogrified the Canadian winter. Only in 1970, after millions of Canadians were being exposed to this new form of noise, was the National Research Council able to conduct research demonstrating that existing machines "present a definite hazard to hearing." Their report showed that machines then on the market frequently exceeded 110 dBA at the drivers' ear. The NRC recommended reducing the noise to 85 dBA (thereby at least lessening the risk of hearing damage) and they showed how this could be done. But the federal government responded by limiting the noise level of new machines to 82 dBA at 50 feet (i.e., approximately 92 dBA at 15 feet).

The intrusion of snowmobiles has now made deafness and ear disease the largest public health problem in the Canadian Arctic, according to Dr. J. D. Baxter, head of McGill University's Otolaryngological Department. In his 1972 address to the Canadian Otolaryngological Society he pointed out that of 156 adult Eskimos examined in one area, 97 showed a significant hearing loss. The Canadian winter used to be noted for its purity and serenity. It was part of the Canadian mythology. The snowmobile has bitched the myth. Without a myth the nation dies.

> . . . no sound issues from a cloudless quarter of the sky.
>
> Lucretius, *On the Nature of Things,*
> VI, 96

The Big Sound Sewer of the Sky   It would be false to assume that man only became airborne in the twentieth century. In fact, man has always been airborne in his imagination, as the numerous magic carpets of folklore prove. The twentieth century has merely reduced the limitless spaces where the imagination soared to rare altitudes to specific air corridors of no intrinsic significance whatsoever. Listen to the sky. The whirring and scraping against the air is nothing but the wounds of a crippled imagination made audible. At one time it was only those unfortunate enough to live near airports who really suffered from aircraft noise. In those days a passing plane turned all heads upward. But since the Second World War all this has changed.

Sometimes I give a class of students the assignment: "You are facing south. You are to wait until a sound passes you by traveling from northeast to southwest." It may take two minutes. It may take two hours. Usually it takes two minutes. Usually it is a plane. "Air travel is doubling every five years, and air freight is growing still faster. . . . Thus . . . the noise goes up in the ratio of the horsepower used in the industry as a whole, that is, it doubles every five years in aviation."

This forecast refers only to the spread of noise energy in the sky. It assumes that we will continue to employ present-day aircraft but simply in greater numbers. To this we must add the very special problems of supersonic transport or any other aberrations that the international aviation industry may still be perpetrating on the drawing boards.

As every home and office is gradually being situated along the world runway, the aviation industry, perhaps more effectively than any other, is destroying the words "peace and quiet" in every world language. For noise in the sky is distinguished radically from all other forms of noise in that it is not localized or contained. The plangent voice of the airplane motor beams down directly on the whole community, on roof, garden and window, on farm and suburb as well as city center.

In our research on the Vancouver soundscape we showed that the annual traffic of aircraft over a downtown park in 1970 was 23,000 per year and that this had grown by 1973 to 38,700—a trend well in line with the quotation above. We also showed that in 1973 the same park soundscape was filled with aircraft noise, from the time each flight was detected on the acoustic horizon until it disappeared, for an average of 27 minutes per hour; and from our research we are able to predict that if the trend continues the noise will be total and uninterrupted by 1981.

A great deal of research has taken place on aircraft noise and it is going forward today more strenuously than ever; but the problem continues to grow. While most of the research has concentrated on the superscreams of jets (and it has succeeded in making the jumbo jets slightly quieter than their predecessors), the insidious jamming of smaller aircraft—for instance, the bitter-batter of helicopters—has been given practically no attention.

The advent of supersonic transports has succeeded in focusing additional public attention on the problem of aircraft noise. Not only do such aircraft produce more noise on take-off and landing, resulting in "a growth in far afield noise accompanied by a serious worsening of the lateral noise spread in the vicinity of the airport," but the most critical feature of this aircraft is that by flying faster than the speed of sound it produces an additional thunderclap called a sonic boom. Unlike the sound of other aircraft, the bang-zone of the supersonic transport boom is about fifty miles wide and extends along the entire length of the aircraft flight path. Supersonic aircraft turn the whole world into an airport.

Let's use the German word *Überschallknall* instead of sonic boom; its ugly syllabification seems more suitable. In addition to its startling noise, the heavier vibrations of the *Überschallknall* can cause serious property damage, smash windows, crack walls and ceilings. On the basis of trial runs of supersonic aircraft in the U.S.A. (the small fighter variety only) and the resulting damage suits filed, it has been estimated that each supersonic flight across that country would startle up to forty million people. In Chicago, test flights over the city resulted in 6,116 complaints and 2,964 damage claims.

As a result of these forecasts, and because in order for supersonic aircraft to be economically viable they must be flown at supersonic speeds as frequently as possible, the Americans in 1972 abandoned their plans to develop such aircraft for commercial purposes. Many countries of the world have banned the flight of supersonic aircraft over their territories, and while the British and French as well as the Russians have such planes, they are now beginning to look like the biggest white elephants of all time.

The supersonic aircraft was an attempt to outmaneuver sound. It failed.

The Deaf Ear of the Aviator   Rather than assist in finding solutions to the problems of aircraft noise, the commercial airlines have turned a deaf ear. They have preferred instead to spend enormous sums of money to pretend that the problem does not exist. If planes make any sound at all, the advertising implies, they are happy sounds. Witness:

- Eastern Airlines "Whisper Jet Service"
- "Fly the Friendly Skies of United."
- "Trident-Two is fast, smooth, quiet and reliable." (BEA)
- "Fly across the Atlantic on the Quiet." (BOAC)
- "We have smart new DC-9 jets with engines quietly at the rear." (Air Jamaica)
- "The DC-10 is a quiet plane that whispers its way through airports." (KLM)
- "More and more people-pleasing 747s are bringing more and more big-jet comfort to more and more cities and towns." (Boeing)

Big jets as people pleasers? Question: What obligation does an airline have to people outside or beneath its aircraft?

On the Acropolis in Athens there is a sign reading:

THIS IS A SACRED PLACE.
IT IS FORBIDDEN TO SING OR
MAKE LOUD NOISES OF ANY KIND.

When I was last there in 1969 the Acropolis was grazed by seventeen jets. Against this hypocrisy we offer the news that Christ and Buddha were also aviators, and wonder what kind of noise they made as they mounted up into the air.

Counter-Revolution   Opposing the developments described in this chapter, there has been, over the past decade, a counter-revolution in many countries around the world. Technological noise is the target for increasing opposition and in a rapidly growing number of instances it is being met directly by noise abatement legislation. As the dangers of excessive noise have been known for at least one hundred fifty years, this sudden expression of interest in the subject, while welcome, raises the question: *Why only now?* Perhaps it is part of a general criticism of the direction in which reckless technology has been taking us. If this is so, the industrialist as God has fallen, and his divine license to make the Sacred Noises without prosecution has ended. I think, and I am merely testing an idea in this sentence, that what we are witnessing in the recent noise abatement campaigns is not so much an attempt to silence the world as an attempt to wrest Sacred Noise from industry as a prelude to the discovery of a more trustworthy proprietor to whom the power may be bequeathed.

SIX

~~~~~~

The Electric Revolution

The Electric Revolution extended many of the themes of the Industrial Revolution and added some new effects of its own. Owing to the increased transmission speed of electricity, the flat-line effect was extended to give the pitched tone, thus harmonizing the world on center frequencies of 25 and 40, then 50 and 60 cycles per second. Other extensions of trends already noted were the multiplication of sound producers and their imperialistic outsweep by means of amplification.

Two new techniques were introduced: the discovery of packaging and storing techniques for sound and the splitting of sounds from their original contexts—which I call schizophonia. The benefits of the electroacoustic transmission and reproduction of sound are well enough celebrated, but they should not obscure the fact that precisely at the time hi-fi was being engineered, the world soundscape was slipping into an all-time lo-fi condition.

A good many of the fundamental discoveries of the Electric Revolution had already been made by 1850: the electric cell, the storage cell, the dynamo, the electric arc light. The detailed application of these inventions occupied the remainder of the nineteenth century. It was during this period that the electric power station, the telephone, the radio telegraph, the phonograph and the moving picture came into existence. At first their commercial applications were limited. It was not until the improvement of the dynamo by Werner Siemens (1856) and the alternator by Nikola Tesla (1887) that electrical power could become the generating force for the practical development of the discoveries.

One of the first products of the Electric Revolution, Morse's telegraph (1838), unintentionally dramatized the contradiction between discrete and contoured sound which, as I have said, separates slow from fast-paced societies. Morse used the long line of the telegraph wire to transmit messages broken in binary code, which still relied on digital adroitness, thus maintaining in the telegrapher's trained finger a skill that related him to the pianist and the scribe. Because the finger cannot be wiggled fast enough to produce the fused contour of sound, the telegraph ticks and stutters in the same way as its two contemporary inventions, Thurber's typewriter and Gatling's machine gun. As increased mobility and speed in communication continued to be desired, it was inevitable that, together with the act of letter-scratching, the telegraph should give way to the telephone.

The three most revolutionary sound mechanisms of the Electric Revolution were the telephone, the phonograph and the radio. With the telephone and the radio, sound was no longer tied to its original point in space; with the phonograph it was released from its original point in time. The dazzling removal of these restrictions has given modern man an exciting new power which modern technology has continually sought to render more effective.

The soundscape researcher is concerned with changes in perception and behavior. Let us, for instance, point up a couple of observable changes effected by the telephone, the first of the new instruments to be extensively marketed.

The telephone extended intimate listening across wide distances. As it is basically unnatural to be intimate at a distance, it has taken some time for humans to accustom themselves to the idea. Today North Americans raise their voices only on transcontinental or transoceanic calls; Europeans, however, still raise their voices to talk to the next town, and Asians shout at the telephone when talking to someone in the next street.

The capacity of the telephone to interrupt thought is more important, for it has undoubtedly contributed a good share to the abbreviation of written prose and the choppy speech of modern times. For instance, when Schopenhauer writes at the beginning of *The World as Will and Idea* that he wishes us to consider his entire book as one thought, we realize that he is about to make severe demands on himself and his readers. The real depreciation of concentration began after the advent of the telephone. Had Schopenhauer written his book in my office, he would have completed the first sentence and the telephone would have rung. Two thoughts.

The telephone had already been dreamed of when Moses and Zoroaster conversed with God, and the radio as an instrument for the transmission of divine messages was well imagined before that. The phonograph, too, has a long history in the imagination of man, for to catch and preserve the tissue of living sound was an ancient ambition. In Babylonian mythology there are hints of a specially constructed room in one of the ziggurats where whispers stayed forever. There is a similar room (still in existence) in the Ali Qapu in Isfahan, though in its present derelict state it is difficult to know how it was supposed to have worked. Presumably its highly polished walls and floor gave sounds an abnormal reverberation

time. In an ancient Chinese legend a king has a secret black box into which
he speaks his orders, then sends them around his kingdom for his subjects
to carry out, which I gloss to mean that there is *authority* in the magic of
captured sound. With the invention of the telephone by Bell in 1876 and
the phonograph by Charles Cros and Thomas Edison in 1877 the era of
schizophonia was introduced.

Schizophonia   The Greek prefix *schizo* means split, separated; and
phone is Greek for voice. *Schizophonia* refers to the split between an original
sound and its electroacoustical transmission or reproduction. It is another
twentieth-century development.

Originally all sounds were originals. They occurred at one time in one
place only. Sounds were then indissolubly tied to the mechanisms that
produced them. The human voice traveled only as far as one could shout.
Every sound was uncounterfeitable, unique. Sounds bore resemblances to
one another, such as the phonemes which go to make up the repetition of
a word, but they were not identical. Tests have shown that it is physically
impossible for nature's most rational and calculating being to reproduce a
single phoneme in his own name twice in exactly the same manner.

Since the invention of electroacoustical equipment for the transmission and storage of sound, any sound, no matter how tiny, can be blown
up and shot around the world, or packaged on tape or record for the
generations of the future. We have split the sound from the maker of the
sound. Sounds have been torn from their natural sockets and given an
amplified and independent existence. Vocal sound, for instance, is no
longer tied to a hole in the head but is free to issue from anywhere in the
landscape. In the same instant it may issue from millions of holes in
millions of public and private places around the world, or it may be stored
to be reproduced at a later date, perhaps eventually hundreds of years after
it was originally uttered. A record or tape collection may contain items
from widely diverse cultures and historical periods in what would seem,
to a person from any century but our own, a meaningless and surrealistic
juxtaposition.

The desire to dislocate sounds in time and space had been evident for
some time in the history of Western music, so that the recent technological
developments were merely the consequences of aspirations that had already been effectively imagined. The secret *quomodo omnis generis instrumen-
torum Musica in remotissima spacia propagari possit* (whereby all forms of
instrumental music could be transmitted to remote places) was a special
preoccupation of the musician-inventor Athanasius Kircher, who discussed the matter in detail in his *Phonurgia Nova* of 1673. In the practical
sphere, the introduction of dynamics, echo effects, the splitting of resources, the separation of soloist from the ensemble and the incorporation
of instruments with specific referential qualities (horn, anvil, bells, etc.)

were all attempts to create virtual spaces which were larger or different from natural room acoustics; just as the search for exotic folk music and the breaking forward and backward to find new or renew old musical resources represents a desire to transcend the present tense.

When, following the Second World War, the tape recorder made incisions into recorded material possible, any sound object could be cut out and inserted into any new context desired. Most recently, the quadraphonic sound system has made possible a 360-degree soundscape of moving and stationary sound events which allows any sound environment to be simulated in time and space. This provides for the complete portability of acoustic space. Any sonic environment can now become any other sonic environment.

We know that the territorial expansion of post-industrial sounds complemented the imperialistic ambitions of the Western nations. The loudspeaker was also invented by an imperialist, for it responded to the desire to dominate others with one's own sound. As the cry broadcasts distress, the loudspeaker communicates anxiety. "We should not have conquered Germany without . . . the loudspeaker," wrote Hitler in 1938.

I coined the term schizophonia in *The New Soundscape* intending it to be a nervous word. Related to schizophrenia, I wanted it to convey the same sense of aberration and drama. Indeed, the overkill of hi-fi gadgetry not only contributes generously to the lo-fi problem, but it creates a synthetic soundscape in which natural sounds are becoming increasingly unnatural while machine-made substitutes are providing the operative signals directing modern life.

Radio: Extended Acoustic Space   A character in one of Jorge

Luis Borges's stories dreads mirrors because they multiply men. The same might be said of radios. By 1969, Americans were listening to 268,000,000 radios, that is, about one per citizen. Modern life has been ventriloquized. The domination of modern life by the radio did not take place unnoticed; but whereas opposition to the Industrial Revolution had come from the working classes, who feared the loss of their jobs, the principal opponents of the radio and the phonograph were the intellectuals. Emily Carr, who wrote and painted in the British Columbia wilderness, hated the radio when she first heard it in 1936.

> When I go to houses where they are turned on full blast I feel as if I'd go mad. Inexplicable torment all over. I thought I ought to get used to them and one was put in my house on trial this morning. I feel as if bees had swarmed in my nervous system. Nerves all jangling. Such a feeling of angry resentment at that horrid metallic voice. After a second I have to clap it off. Can't stand it. Maybe it's my imperfect hearing? It's one of the wonders of the age, simply marvelous. I know that but I *hate* it.

Hermann Hesse, in *Der Steppenwolf* (1927), was disturbed by the poor fidelity of the new electroacoustical devices for the reproduction of music.

> At once, to my indescribable astonishment and horror, the devilish metal funnel spat out, without more ado, its mixture of bronchial slime and chewed rubber; that noise that possessors of gramophones and radio sets are prevailed upon to call music. And behind the slime and the croaking there was, sure enough, like an old master beneath a layer of dirt, the noble outline of that divine music. I could distinguish the majestic structure and the deep wide breath and the full broad bowing of the strings.

But more than this, Hesse was revolted by the schizophonic incongruities of broadcasting.

> It takes hold of some music played where you please, without distinction or discretion, lamentably distorted, to boot, and chucks it into space to land where it has no business to be. . . . When you listen to radio you are a witness of the everlasting war between idea and appearance, between time and eternity, between the human and the divine . . . radio . . . projects the most lovely music without regard into the most impossible places, into snug drawing-rooms and attics and into the midst of chattering, guzzling, yawning and sleeping listeners, and exactly as it strips this music of its sensuous beauty, spoils and scratches and beslimes it and yet cannot altogether destroy its spirit.

Radio extended the outreach of sound to produce greatly expanded profiles, which were remarkable also because they formed interrupted acoustic spaces. Never before had sound disappeared across space to reappear again at a distance. The community, which had previously been defined by its bell or temple gong, was now defined by its local transmitter.

The Nazis were the first to use radio in the interests of totalitarianism, but they have not been the last; and little by little, in both East and West, radio has been employed more ruthlessly in culture-molding. Readers of Solzhenitsyn's novel *Cancer Ward* will recall the "constant yawping" of the radio which greeted Vadim when he went to the hospital and the way he detested it. I recall, twenty years ago, hearing the same loudspeakers blaring out their cacophonies of patriotism and spleen on station platforms and in public squares throughout Eastern Europe. But broadcasting has now gone public in the West as well. It may be hard for younger readers to appreciate what has happened but, up until about a decade ago, one of the most salient differences between cities like London or Paris and Bucharest or Mexico City was that in the former there were no radios or music in public places, restaurants or shops. In those days, particularly during the summer months, BBC announcers would regularly request listeners to keep their radios at a low volume in order not to disturb the neighborhood. In a dramatic reversal of style, British Railways recently began beaming the BBC regional service throughout railway stations (I have heard it over loudspeakers in Brighton Railway Station, 1975). But they still have a long way to go to catch Australian Railways, which plays the ABC light program on trains from 7 a.m to 11 p.m. during the three-day run from Sydney to Perth. In my compartment in 1973 it was impossible to shut it off.

In the early days one listened to the radio selectively by studying the program schedule, but today programs are overlooked and are merely overheard. This change of habit prepared modern society to tolerate the walls of sound with which human engineering now orchestrates the modern environment.

The radio was the first sound wall, enclosing the individual with the familiar and excluding the enemy. In this sense it is related to the castle garden of the Middle Ages which, with its birds and fountains, contradicted the hostile environment of forest and wilderness. The radio has actually become the bird-song of modern life, the "natural" soundscape, excluding the inimical forces from outside. To serve this function sound need not be elaborately presented, any more than wallpaper has to be painted by Michelangelo to render the drawing room attractive. Thus, the development of greater fidelity in sound reproduction, which occupied the first half of the present century—and in a way may be thought of as analogous to the development of oil paints, which also rendered possible greater veracity in art—is now canceled by a tendency to return to simpler forms of expression. For instance, while the transition from mechanical to electrical recording (Harrison and Maxfield) extended the available bandwidth from three to seven octaves, the transistor radio reduced it again to something like its former state. The habit of listening to transistor radios outdoors in the presence of additional ambient noise, often in circumstances which reduce the signal-to-noise ratio to approximately one to one, has in turn suggested the inclusion of additional noise which, in some popular music, is now engineered right onto the disc, often in the form of electroacoustical feedback. This, in turn, leads to new evaluations of what is signal and what is noise in the whole constantly changing field of aural perception.

The Shapes of Broadcasting   Radio programing needs to be analyzed in as much detail as an epic poem or musical composition, for in its themes and rhythms will be found the pulse of life. But detailed studies of this kind appear never to have been undertaken. The structural principles of such an analytical undertaking will be developed in the Rhythm and Tempo chapter of Part Four, but it will not be out of place here to make a few general comments.

At first radio broadcasts were isolated presentations, surrounded by extended (silent) station breaks. This occasional approach to broadcasting, now absent from domestic radio, can still to some extent be experienced with shortwave broadcasts, where station breaks are often several minutes long and are accompanied by short musical phrases or signature tunes.

(This attractive practice is only slightly spoiled by the unlikely choice of instruments used on some stations: thus, the calls of Jordan and Kuwait are played on the clarinet, those of Jamaica and Iran on the vibraphone— that is, they are played on instruments so distinctly non-indigenous that one might suppose they were originally recorded in New York.)

During the 1930s and 1940s schedules were filled out until the whole day was looped together in unsettled connectivity. The modern radio schedule, a confection of material from various sources, joined in thought-ful, funny, ironic, absurd or provocative juxtapositions, has introduced many contradictions into modern life and has perhaps contributed more than anything else to the breakup of unified cultural systems and values. It is for this reason that the study of joins in broadcasting is of great importance. The montage was first employed in film because it was the first art form to be cut and spliced; but since the invention of magnetic tape and the compression of the schedule, the shapes of broadcasting have followed the editor's scissors also.

The function of the montage is to make one plus one equal three. The film producer Eisenstein—one of the first to experiment with montage— defines the effect as consisting "in the fact that two film pieces of any kind, placed together, inevitably combine into a new concept, a new quality, arising out of that juxtaposition." The *non sequiturs* of the montage may be incomprehensible to the innocent though they are easily accommodated by the initiated. I recall one night in Chicago, at the height of the Vietnam War, listening to an on-the-spot report of the grisly affair, sponsored by Wrigley's Chewing Gum, whose jingle at the time was "Chew your little cares away!" I mentioned the experience to a class of students at North-western University the next day. They were interested in my opposition to the war, but failed to see my point about the gum. For them the elements had been montaged as part of a way of life.

Since the advent of the singing commercial on North American radio, popular music and advertising have formed the main material of the radio montage, so that today, by means of quick cross-fades, direct cuts or "music under" techniques, songs and commercials follow one another in quick and smooth succession, producing a commercial life style that is entertaining ("buy baubles for your bippy") and musical entertainment that is profitable ("five million sold").

Radio introduced the surrealistic soundscape, but other electroacous-tical devices have had an influence in rendering it acceptable. The record collection, which one may observe in almost every house of the civilized world, is often equally eclectic and bizarre, containing stray items from different periods or countries, all of which may nevertheless be stacked on the same phonograph for successive replay.

I am trying to illustrate the irrationality of electroacoustic juxtaposi-tioning in order that it might cease to be taken for granted. One last story. A friend was once on an aircraft that supplies a selection of recorded programs of different types for earphone listening. Choosing the program of classical music he settled back in his seat to listen to Wagner's *Meister-*

singer. As the overture soared to a climax, the disturbed voice of the stewardess suddenly interrupted the music to announce: "Ladies and gentlemen, the toilets are plugged up and must be flushed with a glass of water."

As the format of radio tightened, its tempo increased, substituting superficiality for prolonged acts of concentration. Heavyweight fare like the famous BBC Third Programme was dismissed to be replaced by material with more twist and appeal. Each station and each country has its own tempo of broadcasting, but in general it has been speeded up over the years, and its tone is moving from the sedate toward the slaphappy. (I am speaking here only of Western-style broadcasting; I am not sufficiently familiar with the monolithic cultures of Russia or China.) In the West, material is being increasingly pushed together, overlapped. In a World Soundscape Project in 1973 we counted the number of separate items on four Vancouver radio stations over a typical eighteen-hour day. Each item (announcement, commercial, weather report, etc.) represented a change of focus. The results ran as follows:

STATION	TOTAL NUMBER OF ITEMS	HOURLY AVERAGE
CBU	635	35.5
CHQM	745	41.0
CJOR	996	55.5
CKLG	1097	61.0

Stations broadcasting popular music are the fastest-paced. The duration of individual items of any kind rarely exceeds three minutes on North American pop stations. Here the recording industry discloses a secret. On the old ten-inch shellac disc, the recording duration was limited to slightly over three minutes. As this was the first vehicle for popular music, all pop songs were abbreviated to meet this technical limitation. But curiously, when the long-play disc was introduced in 1948, the length of the average pop song did not increase in proportion. This suggests that some mysterious law concerning average attention span may have been inadvertently discovered by the older technology.

One acoustic effect is rarely heard on North American radios: silence. Only occasionally, during broadcasts of theater or classical music, do quiet and silence achieve their full potentiality. A graphic level recording of a popular station will show how the program material is made to ride at the maximum permissible level, a technique known as compression because the available dynamic range is compressed into very narrow limits. Such broadcasting shows no dynamic shadings or phrasing. It does not rest. It does not breathe. It has become a sound wall.

Sound Walls   Walls used to exist to delimit physical and acoustic space, to isolate private areas visually and to screen out acoustic interferences. Often this second function is unstressed, particularly in modern buildings. Confronted with this situation modern man has discovered what might be called *audioanalgesia,* that is, the use of sound as a painkiller, a distraction to dispel distractions. The use of audioanalgesia extends in modern life from its original use in the dental chair to wired background music in hotels, offices, restaurants and many other public and private places. Air-conditioners, which produce a continuous band of pink noise, are also instruments of audioanalgesia. It is important in this respect to realize that such masking sounds are not intended to be listened to consciously. Thus, the Moozak industry deliberately chooses music that is nobody's favorite and subjects it to unvenomed and innocuous orchestrations in order to produce a wraparound of "pretty," designed to mask unpleasant distractions in a manner that corresponds to the attractive packages of modern merchandising to disguise frequently cheesy contents.

Walls used to exist to isolate sounds. Today sound walls exist to isolate. In the same way the intense amplification of popular music does not stimulate sociability so much as it expresses the desire to experience individuation ... aloneness ... disengagement. For modern man, the sound wall has become as much a fact as the wall in space. The teenager lives in the continual presence of his radio, the housewife in the presence of her television set, the worker in the presence of engineered music systems designed to increase production. From Nova Scotia comes word of the continuous use of background music in school classrooms. The principal is pleased with the results and pronounces the experiment a success. From Sacramento, California, comes news of another unusual development: a library wired for rock music in which patrons are encouraged to talk. On the walls are signs stating NO SILENCE. The result: circulation, especially among the young, is up.

> They never sup without music; and there is always fruit served up after the meat; while they are at table, some burn perfumes and sprinkle about fragrant ointments and sweet waters: in short, they want nothing that may cheer their spirits.
> Sir Thomas More, *Utopia*

Moozak   If the Christmas card angels offer any proof, utopian creatures are forever smiling. Thus Moozak, the sound wall of paradise, never weeps. It is the honeyed antidote to hell on earth. Moozak starts out with the high motive of orchestrating paradise (it is often present in writings about utopias) but it always ends up as the embalming fluid of earthly boredom. It is natural then that the testing-ground for the Moozak industry should have been the U.S.A., with its highly idealistic Constitution and the cruddy realities of its modern life styles. The service pages of the telephone directories beam out its advertisements to clients in every North American city.

MUZAK IS MORE THAN MUSIC—PSYCHOLOGICALLY PLANNED—FOR TIME AND PLACE
—JUST FLIP THE SWITCH—NO MACHINES TO ATTEND/FRESH PROGRAMS EACH DAY
—NO REPETITION—ADVISED BY BOARD OF SCIENTIFIC ADVISORS—OVER 30 YEARS OF
RESEARCH—PAGING AND SOUND SERVICE—FAST ROUND-THE-CLOCK SERVICE—
MUZAK BRAND EQUIPMENT—OFFICES—INDUSTRIAL PLANTS—BANKS—HOSPITALS—
RETAIL STORES—HOTELS AND MOTELS—RESTAURANTS—PROFESSIONAL OFFICES—
*SPECIALISTS IN THE PSYCHOLOGICAL AND PHYSIOLOGICAL APPLICATIONS OF
MUSIC.*

Facts on Moozak program design are elementary. The programs are selected and put together in several American cities for mass distribution.
"... program specialists ... assign values to the elements in a musical
recording, i.e., tempo (number of beats per minute); rhythm (waltz, fox
trot, march); instrumentation (brass, woodwinds, strings), and orchestra
size (5 piece combo, 30 piece symphony, etc.)." There are few solo vocalists
or instrumentalists to distract the listener. The same programs are played
to both people and cows, but despite the happy claim that production has
in both cases been increased, neither animal seems yet to have been elevated into the Elysian Fields. While the programs are constructed to give
what the advertising calls "a progression of time"—that is, the illusion that
time is dynamically and significantly passing—the implicit malaise behind
the claim is that for most people time continues to hang heavily. "Each
15-minute segment of MUZAK contains a rising stimulus which provides
a logical sense of forward movement. This affects boredom or monotony
and fatigue."

Although no precise growth statistics have ever been published, there
can be no doubt that these bovine sound slicks are spreading. This does
not perhaps so much indicate a lack of public interest in silence as it
demonstrates that there is more profit to be made out of sound, for another
claim of the Mooze industry is that it provides a "relaxed background to
profit." When we interviewed 108 consumers and 25 employees in a Vancouver shopping mall, we discovered that while only 25 percent of the
shoppers thought they spent more as a result of the background music, 60
percent of the employees thought they did.

Against the slop and spawn of Moozak and broadcast music in public
places a wave of protest is now clearly discernible. Most notable is a
resolution unanimously passed by the General Assembly of the International Music Council of UNESCO in Paris in October, 1969.

We denounce unanimously the intolerable infringement of individual
freedom and of the right of everyone to silence, because of the abusive
use, in private and public places, of recorded or broadcast music. We
ask the Executive Committee of the International Music Council to
initiate a study from all angles—medical, scientific and juridical—
without overlooking its artistic and educational aspects, and with a
view to proposing to UNESCO, and to the proper authorities everywhere, measures calculated to put an end to this abuse.

There is a parallel to this resolution: when, in 1864, Michael Bass proposed his Bill to prohibit the sounds of street singing in the city of London, he drew substantial support from the musical profession itself. With the 1969 UNESCO resolution sonic overkill was apprehended by the musicians of the world as a serious problem. For the first time in history an international organization involved primarily with the *production* of sounds suddenly turned its attention to their *reduction*. In *The New Soundscape* I had already warned music educators that they would now have to be as concerned about the prevention of sounds as about their creation, and I suggested that they should join noise abatement societies to familiarize themselves with this new theme for the music room.

In any historical study of the soundscape, the researcher will repeatedly be struck by shifts in the perceptual habits of a society, instances where the figure and the ground exchange roles. The case of Moozak is one such instance. Throughout history music has existed as figure—a desirable collection of sounds to which the listener gives special attention. Moozak reduces music to ground. It is a deliberate concession to lo-fi-ism. It multiplies sounds. It reduces a sacred art to a slobber. Moozak is music that is not to be listened to.

By creating a fuss about sounds we snap them back into focus as figures. The way to defeat Moozak is, therefore, quite simple: listen to it.

Moozak resulted from the abuse of the radio. The abuse of Moozak has suggested another type of sound wall which is now rapidly becoming a fixture in all modern buildings: the screen of white noise, or as its proponents prefer to call it "acoustic perfume." The hiss of the air-conditioner and the roar of the furnace have been exploited by the acoustical engineering profession to mask distracting sound, and where they are in themselves not sufficiently loud, they have been augmented by the installation of white noise generators. A desideratum from America's most prominent firm of acoustical engineers to the head of a music department shows us that if music can be used to mask noise, noise can also be used to mask music. It ran: "Music Library: There should be enough mechanical noise to mask page turning and foot movement sounds." The mask hides the face. Sound walls hide characteristic soundscapes under fictions.

Prime Unity or Tonal Center   In the Indian *anāhata* and in the Western Music of the Spheres man has constantly sought some prime unity, some central sound against which all other vibrations may be measured. In diatonic or modal music it is the fundamental or tonic of the mode or scale that binds all other sounds into relationship. In China an artificial center of gravity was created in 239 B.C. when the Bureau of Weights and Measures established the Yellow Bell or Huang Chung as the tone from which all others were measured.

It is, however, only in the electronic age that international tonal centers have been achieved; in countries operating on an alternating current of 60 cycles, it is this sound which now provides the resonant frequency, for it will be heard (together with its harmonics) in the operation of all electrical devices from lights and amplifiers to generators. Where C is tuned to 256 cycles, this resonant frequency is B natural. In ear training exercises I have discovered that students find B natural much the easiest pitch to retain and to recall spontaneously. Also during meditation exercises, after the whole body has been relaxed and students are asked to sing the tone of "prime unity"—the tone which seems to arise naturally from the center of their being—B natural is more frequent than any other. I have also experimented with this in Europe where the resonant electrical frequency of 50 cycles is approximately G sharp. At the Stuttgart Music High School I led a group of students in a series of relaxation exercises and then asked them to hum the tone of "prime unity." They centered on G sharp.

Electrical equipment will often produce resonant harmonics and in a quiet city at night a whole series of steady pitches may be heard from street lighting, signs or generators. When we were studying the soundscape of the Swedish village of Skruv in 1975, we encountered a large number of these and plotted their profiles and pitches on a map. We were surprised to find that together they produced a G-sharp major triad, which the F-sharp whistles of passing trains turned into a dominant seventh chord. As we moved about the streets on quiet evenings, the town played melodies.

The Electric Revolution has thus given us new tonal centers of prime unity against which all other sounds are now balanced. Like mobiles, whose movements may be measured from the string on which they are suspended, the sound mobiles of the modern world are now interpretable by means of the thin line fixture of the operating electrical current.

To relate all sounds to one sound that is continuously sounding (i.e., a drone) is a special way of listening. In respect to this development there is an interesting feature of Indian music which might bear further investigation in terms of its relevance for young people growing up in the electronic culture of today. Alain Daniélou explains:

> The modal group of musical systems, to which practically all Indian music belongs, is based on the establishment of relations between a permanent sound fixed and invariable . . . the [drone], and successive sounds, the notes. . . . Indian music . . . is built on the independent relationship of each note to the tonic. The relationship to the tonic determines the meaning of any given sound. The tonic must therefore be constantly heard.

Could this account for the recent popularity of Indian music among the young of the West? One of the key words in the vocabulary of young Americans during the early seventies was "vibrations," i.e., a cosmic sound giving prime unity, a concentration or gathering point from which all other sounds are perceived tangentially.

Interlude

SEVEN

~~~~~~

Music, the Soundscape and Changing Perceptions

Throughout the first two parts of this book I have made numerous references to music. In this interlude between soundscape description and soundscape analysis I want to examine the relationship between music and the soundscape in greater detail. Music forms the best permanent record of past sounds, so it will be useful as a guide to studying shifts in aural habits and perceptions. Europe has been the most dynamic continent over the past five hundred years, so it is in the shapes of European music that these changes can best be measured—at least until America begins to provide the dominating cultural influence in the twentieth century. This is a theme that has remained little explored, for historians and analysts have concentrated on showing how musicians have drawn forth music from the imagination or from other forms of music. But musicians also live in the real world and in various discernible ways the sounds and rhythms of different epochs and cultures have affected their work, both consciously and unconsciously.

Music is of two kinds: absolute and programmatic. In absolute music, composers fashion ideal soundscapes of the mind. Programmatic music is imitative of the environment and, as its name indicates, it can be paraphrased verbally in the concert program. Absolute music is disengaged from the external environment and its highest forms (the sonata, the quartet, the symphony) are conceived for indoor performance. Indeed, they seem to gain importance in direct ratio to man's disenchantment with the external soundscape. Music moves into concert halls when it can no longer be effectively heard out of doors. There, behind padded walls, concentrated listening becomes possible. That is to say, the string quartet and urban pandemonium are historically contemporaneous.

The Concert Hall as a Substitute for Outdoor Life

The concert hall simultaneously brought about absolute musical expression and also the most decisive imitations of nature. The conscientious imitation of landscape in music corresponds historically to the development of landscape painting, which seems to have been first cultivated by the Flemish painters of the Renaissance and developed into the principal genre of painting in the nineteenth century. Such developments are explicable only as a result of the displacement of the art gallery farther and farther from the natural landscape in the hearts of growing cities. Imitations of nature were then created to be exhibited in unnatural settings. Here they functioned as so many windows, releasing the spectator onto different scenes. An art gallery is a room with a thousand avenues of departure, so that once having entered, one loses the door back to the real world and must go on exploring. In the same way, a descriptive piece of music turns the walls of the concert hall into windows, exposed to the country. By means of this metaphorical fenestration we break out of the confinements of the city to the free *paysage* beyond.

This is certainly true of the nature descriptions of eighteenth-century composers such as Vivaldi, Handel or Haydn. Their landscapes are well populated with birds, animals and pastoral people—shepherds, villagers, hunters. Their descriptions are colorful, exact and benign. The music of Haydn is certainly not bereft of drama, but it is a music of happy endings, as we observe in *The Seasons,* where, following the storm, the clouds part to reveal the setting sun, while the cattle turn refreshed to the stable, the curfew bell sounds (the measures of the orchestration suggest that it is eight o'clock) and the world turns to that "soothing sleep that guileless heart and goodly health" ensures. For Haydn nature is the grand provider, and the pastoral people of his tableau enjoy an "easy and insatiable exploitation of the land and its creatures."

Given the differences in style, Handel's landscapes are close in tone to those of Haydn. In a work like *L'Allegro ed il Penseroso,* adapted from Milton's famous duet of poems, we are presented with all the familiar features (birds, gently rolling countryside, hounds and horns) but in one of the arias, for baritone and chorus, there is an uncommon description to the words

> Populous cities please me then,
> And the busy hum of men,

for here oboes, trumpets and kettledrums join the orchestra and chorus in a rousing tribute to metropolitan life. Living in the city, Handel was one of the first composers to be influenced by the bustle of urban activity and is said to have derived inspiration from the singing and noises in the streets. Although he possessed an orthodox musical talent for nature description, there is nothing in musical literature to compare with Handel's ear for urban acoustics until we approach the scores of Berlioz and Wagner.

The landscapes of Handel and Haydn are as rich in detail as the paintings of Breughel and they are just as carefully structured. Michelangelo had criticized the Flemish painters for failing to exercise selection in their subject matter; instead of focusing on one thing they included everything in view. Indeed, the compositions to which I have alluded share a similar feature, for they are wide-angle tableaux; the composer observes the landscape at a distance. Nature performs and he provides the secretarial services.

Only in the landscapes of the romantic era does the composer intrude to color nature with his own personality or moods. Natural events are then made to synchronize or to compete ironically with the moods of the artist. I have already mentioned how this technique of sympathetic vibration originated in pastoral poetry (Theocritus, Virgil), where it came to be known by literary critics as "the pathetic fallacy," but we do not encounter effective employment of the technique in music history until the song cycles of Schubert and Schumann.

Schubert has often made the landscape perform for him. In a song such as "Der Lindenbaum" ("The Lime Tree") from *Die Winterreise,* the moods of the poet-composer stimulate the tree, causing its branches to move gently (summer) or violently (winter), while day and night thoughts are distinguished by major and minor tonalities. In Schumann's *Dichterliebe,* the landscape maintains its gay summer colors while the poet's joy turns to grief, a bitterly ironical situation which is fully exploited in contrasts between vocalist and pianist.

Throughout the history of Western music, the sounds of nature (particularly those of wind and water) have been frequently and adequately rendered, as have bells, birds, firearms and hunting horns. We have already touched on street cries and have also mentioned the suggestibility of the solo woodwind instrument for the pastoral landscape. Let us inspect a few of the others.

Music, Bird-Song and Battlefields   Bird-song in music has a parallel in the enclosed garden of literature. Before the landscape of Europe was cultivated, nature presented a vast and fearful spectacle. The medieval garden was an attempt to create a benign and flowering place where love, human and divine, could be fulfilled. Thus, in the Cave of Lovers from Gottfried of Strassburg's *Tristan:*

> At their due times you could hear the sweet singing of the birds. Their music was so lovely—even lovelier here than elsewhere. Both eye and ear found their pasture and delight there: the eye its pasture, the ear its delight. There were shade and sunshine, air and breezes, both soft and gentle. . . .
>
> The service they received was the song of the birds, of the lovely, slender nightingale, the thrush and blackbird, and other birds of the

forest. Siskin and calander-lark vied in eager rivalry to see who could give the best service. These followers served their ears and sense unendingly. Their high feast was Love, who gilded all their joys.

Birds contributed to the felicitous atmosphere of the garden, and they were deliberately attracted there by means of feeding and fountains. In Persian gardens, birds had been retained in huge nets. It may well be that the peculiar value set on gardens in the later Middle Ages is a legacy of the Crusaders, who also appear to have brought back the arts of lyric poetry and song from the Middle East. It was in this diminutive meadow then, behind the protective wall of the castle, that the troubadour art flourished, and the voices of the birds were often woven into their songs. It is the same pleasant and docile atmosphere that Nicolas Gombert and Clément Janequin extended in their *Chants des Oiseaux.* Bird-song will always suggest this delicacy of sentiment and I would go so far as to suggest that it appears in music in deliberate contradiction to the brutalities and accidents of external life. It is this way that it enters in opposition to the malignant forces of Wagner's *Ring,* and it is sustained by Olivier Messiaen in our time for the same reason.

The case of firearms is in opposition to this. The cannon was first effectively employed by Edward III of England at Crécy in 1346 and again during the siege of Calais in 1347; but the first full-fledged musical treatment of firearms seems to have been Janequin's vocal exhibition piece *La Bataille de Marignan* of 1545.

> *Frerelelelan fan farirarirarirariri*—went the trumpets
> *Von von von patipatos pon pon pon*—went the cannons.

The effect must always have been comical, with the result that instrumental versions of battle scenes quickly took precedence over vocal in examples too numerous to mention, until we arrive at Beethoven's Battle Symphony, where the imitative gunfire is replaced by the real thing—another sign of Beethoven's evident pugilism.

The devices of program music transform the real space of the concert hall into a garden, a pasture, a forest or a battlefield. These metaphorical spaces gain and lose in popularity over the years, and a study of this subject would give us a good idea of the changing attitudes of urban man to the landscape. To illustrate I will take just one theme, one I have already introduced: the hunting horn. We can now follow its symbolic transformations over a critical period, from the end of the eighteenth century to the beginning of the twentieth.

The Hunting Horn Explodes the Walls of the Concert Hall to Reintroduce the Countryside

Hunting horn motifs were employed for colorful effects in numerous symphonic works during the eighteenth century. Haydn's *La Chasse* symphony (No. 73) is a good example. The robust tones of the horns at once cut through the other instruments to suggest the spirit of outdoor life. "Hark!" sings the chorus in Haydn's *Creation,* as the horns are sounded,

> The clamorous noise that through the
> wood is ringing!
> How clear the shrilling of horns resounds!
> How eager the hounds are all baying!
> Now speeds the fear-rous'd stag: they
> follow, the pack and the hunters too.

The famous huntsmen's chorus from Weber's *Freischütz* expresses the same passion for outdoor life and the high, free spirits of the hunt.

Once the horn has taken on a clear symbolic function, ironic transformations can be played over it. Thus, in Weber's *Oberon,* the opening three notes on the solo horn—one of the most evocative effects in all music—transport us to the wonderful, perfumed gardens of the Orient. The hunting horn has become a magic horn, capable of moving the audience beyond local fields to distant *pays de chimères.*

In the symphonies of Brahms and Bruckner there is, if I am not suffering from a distended imagination, a perceptible transformation of the hunting horn into what we might call the horn of authority, for a certain hectoring quality, almost a stubbornness, is evident here which is absent from the spirited treatment given it by earlier composers.

We have another example of irony in Schubert's "Die Post" from *Die Winterreise,* where the distant post horn dances across the acoustic horizon in the piano accompaniment, while the singer's joy of anticipation turns to melancholy when he realizes that he will receive no letter from his beloved.

More than any instrument, the horn symbolizes freedom and love of the outdoors. When sounded in the concert hall, it collapses the walls and transports us again to the unrestricted spaces of the country. For those who were accustomed to hearing horns regularly, just beyond the city wall, this effect must have been immediately appealing. The horn of freedom achieves heroic proportions in Wagner's *Siegfried,* where it becomes the acoustic symbol of the hero who will one day bring about the collapse of a moribund civilization.

But the most interesting transformation of the horn for our purposes is the last, the one we might call the horn of memory. The most eloquent examples of this transformation are to be found in the symphonies of Gustav Mahler. Already we have it clearly indicated in the opening movement of Mahler's first symphony of 1888. Here hunting horn motifs are suggested first, distantly, on the clarinets, then by offstage trumpets, and finally, very slowly, very nostalgically, on the horns themselves. As the movement gathers impetuously toward its climax, the horns in glissando rips break away in a very rage for freedom. But it is the floating and sentimental quality of Mahler's distant horns that are most memorable, for they prefigured the transmogrification of the landscape itself. Today there is no open countryside left in Europe; there are only fences and parks.

The Orchestra and the Factory   If the solo flute and the hunting horn reflected the pastoral soundscape, the orchestra reflects the thicker densities of city life. From the earliest days the orchestra had shown a tendency to grow in size, but it was not until the nineteenth century that its forces were co-ordinated and its instruments strengthened and scientifically calibrated to give it the complex and powerful sound-producing capabilities which, in terms of intensity alone, made it a competitor with the polynoise of the industrial factory. But there were even greater parallels between the orchestra and the factory, as Lewis Mumford explains:

> ... with the increase in the number of instruments, the division of labor within the orchestra corresponded to that of the factory: the division of the process itself became noticeable in the newer symphonies. The leader was the superintendent and production manager, in charge of the manufacture and assembly of the product, namely the piece of music, while the composer corresponded to the inventor, engineer, and designer, who had to calculate on paper, with the aid of such minor instruments as the piano, the nature of the ultimate product—working out its last detail before a single step was made in the factory. For difficult compositions, new instruments were sometimes invented or old ones resurrected; but in the orchestra the collective efficiency, the collective harmony, the functional division of labor, the loyal cooperative interplay between the leaders and the led, produced a collective unison greater than that which was achieved, in all probability, within any single factory. For one thing, the rhythm was more subtle; and the timing of the successive operations was perfected in the symphony orchestra long before anything like the same efficient routine came about in the factory.
>
> Here, then, in the constitution of the orchestra, was the ideal pattern of the new society. It was achieved in art before it was approached in technics.... Tempo, rhythm, tone, harmony, melody, polyphony, counterpoint, even dissonance and atonality, were all utilized freely to create a new ideal world, where the tragic destiny, the dim longings, the heroic destinies of men could be entertained once more. Cramped by its new pragmatic routines, driven from the marketplace and the factory, the human spirit rose to a new supremacy in the concert hall. Its greatest structures were built of sound and vanished in the act of being produced. If only a small part of the population listened to these works of art or had any insight into their meaning, they nevertheless had at least a glimpse of another heaven

than Coketown's. The music gave more solid nourishment and warmth than Coketown's spoiled and adulterated foods, its shoddy clothes, its jerrybuilt houses.

The orchestra was then an idealization of the aspirations of the nineteenth century, a model which the industrialist-kings tried to emulate in their factory routines.

Even the musical forms cultivated by the nineteenth century seem to have an imperialistic bias: thus, in the first-movement form of the symphony, home base is established (exposition), colonies are developed (*Durchführung*) and the empire is consolidated (recapitulation and coda). It was during this period, too, that the bass bars of all stringed instruments were carefully replaced to produce greater volume of sound; also new brass and percussion instruments were added and the piano replaced the harpsichord, which was no longer strong enough to be heard in the new instrumental consort. The substitution of the punched-string piano for the plucked-string harpsichord typifies the greater aggressiveness of a time in which objects were punched and beaten into existence by means of new industrial processes. Material had once been stroked, carved or kneaded into shape; now it was slugged. The strengthening of the piano, which exchanged sound quality for sound quantity, disturbed the Viennese critic Eduard Hanslick, who realized that amplified music would lead to increased community disturbance.

> Complaints of annoyance caused by neighbourhood pianos are by no means as old as the piano itself. In Mozart's and Haydn's day the piano was a weak, thin box with a soft tone, scarcely audible as far as the front room. The complaints began with the introduction of the piano's stronger tone and extended range, and became a painful outcry only thirty or forty years ago, after the piano manufacturers began in earnest to increase the sound output of the instrument. . . . The full tone and carrying power of the modern piano arise from its great size, its colossal weight and the tensions of its metal-strengthened frame. . . . The instrument has gained this offensive power and offensive character for the first time in our day.

The power these new technical developments permitted was first seized by Beethoven. It would be wrong to think of Beethoven as a product of the Industrial Revolution, which, after all, hardly touched Vienna during his lifetime, but he was certainly an urban composer and his pugilistic temperament made the "offensive" character of the new instruments specially appealing for him, as can immediately be sensed in touching or hearing the opening notes of a work like the *Hammerklavier Sonata*, Op. 106. In principle there is little difference between Beethoven's attempts to *épater les bourgeois* with full-fisted sforzando effects and that of the modern teenager with his motorcycle. The one is an embryo of the other.

The Meeting of Music and Environment   The imperialism of nineteenth-century music reached its apex in the orchestras of
Wagner and Berlioz, which were specially expanded to make possible a
grandiloquent rhetoric, designed alternately to thrill, exalt and crush
swelling metropolitan audiences. Berlioz's ideal orchestra was to include
120 violins, 16 French horns, 30 harps, 30 pianos, and 53 percussion parts.
Wagner had similar ambitions and ended up with an orchestra that constantly threatened to drown the singers—a problem which caused him
much anxiety. It is in this light that one can appreciate Spengler's criticism
of Wagner's art as a sort which "signifies a concession to the barbarism of
the Megalopolis, the beginning of dissolution sensibly manifested in a
mixture of brutality and refinement."

When the orchestra continued to expand into the twentieth century
it was primarily percussion instruments that were added, that is, nonpitched noisemakers capable of sharp attack and rhythmic vitality. The
pastorale and the nocturne then ceased to exist and were replaced by the
machine music of Honegger's *Pacific 231* (1924), an imitation of a locomotive, Antheil's *Ballet Méchanique* (1926), which employed a number of
airplane propellers, Prokofiev's *Pas d'Acier* (*Dance of Steel*), Mossolov's *Iron
Foundry* and Carlos Chávez's *HP* (*Horsepower*), all dating from 1929. Poets
like Ezra Pound and F. T. Marinetti were also going through machine
periods, as were painters like Léger and the artisans of the Bauhaus. In 1924
Pound had written: "I take it that music is the art most fit to express the
fine quality of machines. Machines are now a part of life, it is proper that
men should feel something about them; there would be something weak
about art if it couldn't deal with this new content."

The anomie of modern city life had already been effectively described
in Satie's deadpan *musique d'ameublement*—the original Moozak. When
Satie designed this entertainment for the intermission of a play at a Paris
art gallery in 1920, he intended that the spectators should move about and
ignore the music, which was to be regarded merely as so much upholstery.
Unfortunately, everyone stopped to listen. Music was then still something
to be prized; it had not yet flipped over to its new function as background
drool; and Satie had to rush about crying, "Parlez! Parlez!"

From our point of view the real revolutionary of the new era was the
Futurist experimenter Luigi Russolo, who invented an orchestra of noisemakers, consisting of buzzers, howlers and other gadgets, calculated to
introduce modern man to the musical potential of the new world about
him. In 1913 Russolo proclaimed the event in his manifesto *The Art of
Noises* (*L'Arte dei Rumori*):

> In antiquity, life was nothing but silence. Noise was really not born
> before the 19th century, with the advent of machinery. Today noise
> reigns supreme over human sensibility. . . . In the pounding atmo-

sphere of great cities as well as in the formerly silent countryside, machines create today such a large number of varied noises that pure sound, with its littleness and its monotony, now fails to arouse any emotion. . . . Let's walk together through a great modern capital, with the ear more attentive than the eye, and we will vary the pleasures of our sensibilities by distinguishing among the gurglings of water, air and gas inside metallic pipes, the rumbling and rattlings of engines breathing with obvious animal spirits, the rising and falling of pistons, the stridency of mechanical saws, the loud jumping of trolleys on their rails, the snapping of whips, the whipping of flags. We will have fun imagining our orchestration of department stores' sliding doors, the hubbub of the crowds, the different roars of railroad stations, iron foundries, textile mills, printing houses, power plants and subways. And we must not forget the very new noises of Modern Warfare.

Russolo's experiments mark a flash-point in the history of aural perception, a reversal of figure and ground, a substitution of garbage for beauty. Marcel Duchamp did the same thing about the same time for the visual arts by exhibiting a urinal. It was outrageous because, instead of perpetuating the picture-window mythology of the traditional art gallery, the public was confronted with a doorframe back to the place they had just left.

When John Cage opened the doors of the concert hall to let the traffic noise mix with his own, he was paying an unacknowledged debt to Russolo. An acknowledged debt was paid him by Pierre Schaeffer during the formative years of the *musique concrète* group in Paris. In the practices of *musique concrète* it became possible to insert any sound from the environment into a composition with tape, while in electronic music the hard-edge sound of the tone generator may be indistinguishable from the police siren or the electric egg-beater.

This blurring of the edges between music and environmental sounds may eventually prove to be the most striking feature of all twentieth-century music. In any case, these developments have inescapable consequences for music education. A musician used to be one who listened with seismographic delicacy in the music room, but who put on ear flaps when he left. If there is a noise pollution problem in the world today it is certainly partly and maybe largely owing to the fact that music educators have failed to give the public a total schooling in soundscape awareness, which has, since 1913, ceased to be divisible into musical and nonmusical kingdoms.

Reactions   Marshall McLuhan somewhere says that man only discovered nature after he had wrecked it. So it was at the very time when the natural soundscape was being overrun, it stimulated a whole wave of sensitive reactions in the music of composers as different as Debussy, Ives or Messiaen. There are moments, too, when Bartók's music steams and rustles with all kinds of primordial buzzings suggesting a microcosmic life as close to the grass as was Goethe's ear when he wrote poetry or is the entomologist's microphone when he records the grasshopper's clicking. Just as the microscope revealed a whole new landscape beyond the human eye, so the microphone in a sense revealed new delights missed by the average ear. As a skilled recordist of folk songs Bartók knew this and the evidence is in his quartets and concertos.

Charles Ives, who "glorified America at the same time as he saw it going to hell" (Henry Brant), also reflected a great deal on the dilemma of disappearing nature. Note his songs on the phonograph and the railroad: very ugly sounds. His song about the Indians goes, "Alas, for them their day is o'er . . . the pale man's axe rings through their woods." Ives's heart was with the landscape and in the village, and his uncompleted Universe Symphony was designed to be performed outdoors on the hills and in the valleys.

Olivier Messiaen, like Ives, is an ecological composer. In his music, man is not the supreme triumph of nature but rather an element in a supreme activity called *life*. How different is the impact of even his large orchestral works, like the Turangalila Symphony—so full of birds and the respirating forest—from other orchestral efforts, typified by Strauss's *Ein Heldenleben* (*The Life of a Hero*). How very different, too, is this music from Respighi's *Pines of Rome,* where, for the first time, recorded music (bird-song) was coupled with the symphony orchestra. That was in 1924. Two years earlier Paul Klee had celebrated the mechanical bird in his satirical painting *The Twittering Machine.*

Perhaps the retreat from the commotion of city life had already begun in the nineteenth century (remember that Mahler composed in the country) so that the physical separation of the artist from his public had much to do with his eventual social alienation; but we must turn at this point and give some examples of the interaction between art and the new technology.

Interactions   Throughout the history of soundmaking, music and the environment have bequeathed numerous effects to one another, and the modern era provides striking examples. For instance, while the internal combustion engine gave music the long line of low-information sound, music gave the automobile industry the pitched horn, tuned (in North America) to the major or minor third.*

The development of the Alberti bass of the eighteenth century from galloping horses is another example of environmental influence on art. Consider, for instance, two composers, one living in that century and one living in our own. The former travels everywhere in a carriage. He can't get horses' hooves off his brain and his tunes all go clippety-clop to the opera shop. The latter travels everywhere in his sports car. His music is remarkable for its drones, clusters and whirring effects. Penderecki's mu-

---

*A note of luxury is sounded in the three-tone horn, for it is standard equipment only on the most expensive cars: Cadillac and Lincoln Continental.

sic, for example, leaves the impression that it was conceived somewhere between the airstrip and the Autobahn—I am not criticizing, just pinning down a fact.*

It ought to be obvious also that one of the latest enthusiasms in modern music, phase shifting, has its origin in the machine, or more especially in the machine that employs belts as well as cogs. The cogged machine produces an unvarying clatter, but wherever belts are employed there is slippage giving rise to gradual rhythmic transformations or phase shifts. Such types of machines have been around for some time (the prairie combine is a good example) and they have no doubt infected the minds of numerous young composers who are now engaged in transposing the effect into music. One might argue that the technique was first suggested by the tape recorder rather than the combine, for the first pieces exploiting the effect were composed on tape recorders. No matter, they are both belted machines.

My colleague, Howard Broomfield, also believes that railroads had an important influence on the development of jazz. He claims blue notes (slides from major to minor thirds and sevenths) can be heard in the wail of the old steam whistles. Also the similarity between the clickety-clack of wheels over track ends and the drumbeats (particularly the flam, the ruff and the paradiddle) of jazz and rock music is too obvious to go unnoticed, at least in the clever tape mixes Broomfield has made to prove the point. Since the wheel trucks of different coaches are mounted in different positions (see drawing below), the rhythm of their passage over track ends will vary. By calculating these distances one could notate the precise rhythms produced, and these could be compared with those of different popular bands.

The recording of music on disc and tape has affected composition. All ordered language systems require redundancy. Music is one such system and its redundancy consists in the repetition and recapitulation of principal material. When Mozart repeated a theme six or eight times, it was to help the memory store it for later recall. I do not think it was accidental,

---

*I recently came across an interesting confirmation of the above paragraph in the words of Stockhausen: "I was flying every day for two or three hours over America from one city to the next over a period of six weeks, and my whole time feeling was reversed after about two weeks. I had the feeling that I was visiting the earth and living in the plane. There were just very tiny changes of bluish colour and always this harmonic spectrum of engine noise. At that time, in 1958, most of the planes were propeller planes and I was always leaning my ear—I *love* to fly, I must say—against the window, listening with earphones directly to the inner vibrations. And though theoretically a physicist would have said that the engine sound doesn't change, it changed all the time because I was listening to all the partials within the spectrum. It was a fantastically beautiful experience. And I really discovered the innerness of the engine sounds and watched the slight changes of the blue outside and then the formation of the clouds, this white blanket always below me. I made sketches for *Carré* during that time, and thought I was already very brave in going far beyond the time of memory, which is the crucial time between eight- and sixteen-second-long events." Jonathan Cott, *Stockhausen: Conversations with the Composer*, London, 1974, pp. 30–31.

therefore, that Schoenberg and his followers sought to achieve a musical style which was athematic (i.e., without repetitions and recapitulation) about 1910, at the same time as recording became commercially successful. From then on the recapitulation was on the disc. In fact, the function of the recording industry in providing redundancy and therefore stability in life at a time when the future seems uncertain should not be overlooked, and if the success of radio stations which play the same tunes over and over is any indication, human beings are by no means ignorant of this value. At first it seems paradoxical that in a dynamic and revolutionary era most people should prefer the music of the past, until we realize that for the vast majority of humans today music no longer functions as the antennae of the spirit but as a sensory anchor and stabilizer against future-shock.

Sacred Noise in Search of a New Custodian

Just as the Electric Revolution extended the imperialistic power motive of the Industrial Revolution, but with greater finesse, so the amplifier replaced the orchestra as the ultimate weapon for dominating acoustic space. We have recorded the sound level in the orchestra during a rehearsal of Stravinsky's *Sacre du Printemps* (last section) at 108 decibel peaks, but numerous pop groups have exceeded this volume with a fraction of the manpower. The orchestra ceased growing with the invention of the amplifier—which was first used successfully for a political rally when Woodrow Wilson addressed the League of Nations on September 20, 1919. By this time serious composers had already begun to write works on a smaller scale, works which were especially suitable for the dry acoustics of the broadcasting studio; but popular music, which was frequently performed outdoors, ultimately turned the amplifier into a lethal weapon by pushing sound production up to the threshold of pain. While during the 1960s workmen's compensation boards were introducing limits for noisy industrial environments (85 to 90 decibels is recommended for continuous noise), rock bands were producing peaks of 120 decibels, with the result that when audiologists finally settled down to the task of assessing the damage, they discovered the obvious: rock fans, mostly teenagers, were suffering from "boilermaker's disease."

Now, we will recall that the vibratory effects of high-intensity, low-frequency noise, which have the power to "touch" listeners, had first been experienced in thunder, then in the church, where the bombardon of the organ had made the pews wobble under the Christians, and finally had been transferred to the cacophonies of the eighteenth-century factory. Thus, the "good vibes" of the sixties, which promised an alternative life style, traveled a well-known road, which finally led from Leeds to Liverpool; for what was happening was that the new counterculture, typified by Beatlemania, was actually stealing the Sacred Noise from the camp of the industrialists and setting it up in the hearts and communes of the hippies.

At the Frontiers of Aural Space   We speak of aural space when we plot intensity against frequency on a graph. Time is the third dimension to this space, but for the moment I want to consider the first two in isolation. Aural space is merely a notational convention and should not be confused with acoustic space, which is an expression of the profile of a sound over the landscape. We know that aural space is limited on three sides by thresholds of the audible and on one by a threshold of the bearable. Thus, man may hear sounds from approximately 20 hertz (below which the sense of hearing fuses with that of touch) to 15 or 20 kilohertz, and from zero decibels to approximately 130 decibels (where sound sensation is converted to pain). This is speaking very generally. Actually the shape of aural space is by no means regular, as the outer rim of the graph below indicates.

The growth in intensity of Western music is paralleled by a growth in frequency range. Throughout the past several hundred years new instruments have been designed to push the tonal range out toward the limits of audibility in both directions, until, with contemporary electronic music and hi-fi reproduction equipment, a complete range from approximately 30 hertz to 20,000 hertz is available to the composer and performer. Speaking approximately, we may say that while up to the Renaissance or even up to the eighteenth century, music occupied an area in intensity and frequency range such as that shown in the core of the graph, since that time it has progressively pushed out so that it is practically coincidental with the shape representing the total area of the humanly audible.

Since no sound can be present at all places at once, aural space is to be regarded merely as potential. Within it are set up tensions of opposites. Thus, as the intensity of the modern soundscape or of modern music increases, tranquility diminishes. Sounds are similarly distinguished in their frequency distributions. The popular music we have been considering has, in its choice of instruments, shown a distinct preference for low-

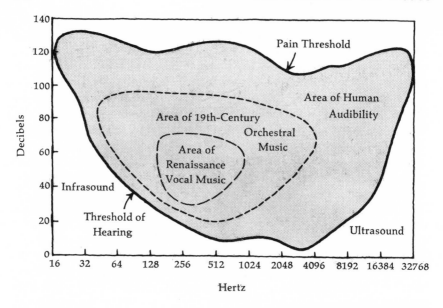

frequency or bass effects, and young people listening to this music generally emphasize this effect by boosting the bass response of their record
players. This is interesting because the longer wavelengths of low-
frequency sounds have more carrying power (as the foghorn demonstrates), and as they are less influenced by diffraction, they are able to
proceed around obstacles and fill space more completely. Localization of
the sound source is more difficult with low-frequency sounds, and music
stressing such sounds is both darker in quality and more directionless in
space. Instead of facing the sound source the listener seems immersed in
it.

Increased Bass Response in Music and the Soundscape

The boost in bass effects in contemporary popular music has its parallel
and has perhaps even received a stimulus from the general increase in
low-frequency environmental sounds. This matter is perceptively discussed by Michel P. Philippot in an article in *New Patterns of Musical
Behaviour.*

> It is recorded, for instance, that in the 17th century the noise in Paris
> was literally unbearable. The same reports inform us about the nature
> of this noise: shouting, carts and carriages, horses, bells, artisans at
> work, etc. From this we may infer that the average sound level must
> have shown marked fluctuations, that its envelope must have had
> peaks and lows so that it was actually "cut up." Besides, the spectrum
> must have been very poor in low frequencies, as all the noises enu-
> merated above belong to the medium and medium-high frequency
> ranges. In the mechanical age and—if we talk of the noise in big cities

—with the invention of the automobile, the noise became more continuous and the lower frequencies were strongly increasing (the deep rumbling of urban traffic, the continuous noise of cars that are driving by, the broad spectrum and long envelope of approaching and departing planes). The "modern" ambient noise might be briefly characterised as heavy and continuous, with slow fluctuations that are difficult to identify and to locate, as this kind of noise tends to encompass us. "I stop talking," said the aged d'Alembert, "when a car drives by." . . . This means that he could still enjoy moments of silence between two cars, a blessing of which the victims of the low and continuous noise in big towns have meanwhile been deprived.

In stressing low-frequency sound popular music seeks blend and diffusion rather than clarity and focus, which had been the aim of previous music and was achieved by separating performers and listeners in counterpoised groups, usually facing one another. As may be suspected, this type of music tends to stress higher frequency sounds to make its directionality clear. Such is the music of the classical concert and its high points are the chamber music of Bach and Mozart. In such music distance is important, and the real space of the concert hall is extended in the virtual space of dynamics—by which effects may be brought into the foreground (*forte*) or allowed to drift back toward the acoustic horizon (*piano*). The formal dress of such concerts helps also to put social space between the participants, for such music belongs to eras of class distinction, to the society of the high- and the lowborn, the master and the apprentice, the virtuoso and the listener. Such music also demands great concentration. This is why silence is observed at concerts where it is performed. Each piece is affectionately placed in a container of silence to make detailed investigation possible.

Thus, the concert hall made concentrated listening possible, just as the art gallery encouraged focused and selective viewing. It was a unique period in the history of listening and it produced the most intellectual music ever created. It contrasts vividly with music designed for outdoor performance, such as folk music, which does not demand great attention to detail, but brings into play what we might call peripheral hearing, similar to the way the eye drifts over an interesting landscape. As the transistor radio revives interest in the outdoor concert and the guitar returns to orchestrate the *baraque de foire* of the rock concert, we may accordingly expect to witness a deterioration of manners at concert halls, as concentrated listening gives way to impressionism.

The Return to the Submarine Home   Another type of listening is produced in the indoor concert from which distance and directionality are absent, i.e., that of much contemporary and popular music as well as that of the living room stereo set. In this case the listener finds himself at the center of the sound; he is massaged by it, flooded by it. Such

listening conditions are those of a classless society, a society seeking unification and integrity. It is by no means a new impulse to seek this kind of
sound space, and in fact it was once beautifully achieved in the singing of
Gregorian chants in the cathedrals of the Middle Ages. The stone walls and
floors of Norman and Gothic cathedrals produced not only an abnormally
long reverberation time (six seconds or more) but also reflected sounds of
low and medium frequencies as well, discriminating against high frequencies above 2,000 hertz owing to the greater absorption of the walls and air
in that range. Anyone who has heard monks chanting plainsong in one of
these old buildings will never forget the effect: the voices seem to issue
from no point but suffuse the building like perfume. In an excellent study
of this subject, the Viennese music sociologist Kurt Blaukopf concluded:

> The sound in Norman and Gothic churches, surrounding the audience,
> strengthens the link between the individual and the community. The
> loss of high frequencies and the resulting impossibility of localising
> the sound makes the believer part of a world of sound. He does not
> face the sound in "enjoyment"—he is wrapped up by it.

The experience of immersion rather than concentration forms one of
the strongest links between modern and medieval man. But we can look
back farther still to determine a common origin. Where then is the dark
and fluid space from which such listening experiences spring? It is the
ocean-womb of our first ancestors: the exaggerated echo and feedback
effects of modern electronic and popular music re-create for us the echoing
vaults, the dark depths of ocean.

Toward the Integrity of Inner Space

Thus, we have a polar-
ity between two types of listening which, to some extent at least, seems
to result from sounds positioned in different frequency bands. We can now
appreciate the dichotomy which seems to separate the nineteenth and
twentieth centuries. Perhaps we can even appreciate McLuhan's claim that
electricity unites men together again.

High frequency	Low frequency
Sound from a distance	Wraparound sound
Perspective	Presence
Dynamics	Sound wall
Orchestra	Electroacoustics
Concentration	Immersion
Air (?)	Ocean-womb

Sound is in more intimate proximity to the listener in the right-hand
column. Let us move the sound source closer still. The ultimate private
acoustic space is produced with headphone listening, for messages received
on earphones are always private property. "Head-space" is a popular expression with the young, referring to the geography of the mind, which

can be reached by no telescope. Drugs and music are the means of invoking entry. In the head-space of earphone listening, the sounds not only circulate around the listener, they literally seem to emanate from points in the cranium itself, as if the archetypes of the unconscious were in conversation. There is a clear resemblance here to the functioning of Nada Yoga in which interiorized sound (vibration) removes the individual from this world and elevates him toward higher spheres of existence. When the yogi recites his mantra, he *feels* the sound surge through his body. His nose rattles. He vibrates with dark, narcotic powers. Similarly, when sound is conducted directly through the skull of the headphone listener, he is no longer regarding events on the acoustic horizon; no longer is he surrounded by a sphere of moving elements. He *is* the sphere. He is the universe.

Headphone listening directs the listener toward a new integrity with himself. But only when he releases the experience by pronouncing the sacred *Om* or singing the Hallelujah Chorus or even the "Star Spangled Banner" does he take his place again with humanity.

PART THREE

~~~~~~~~~~~~~~~

Analysis

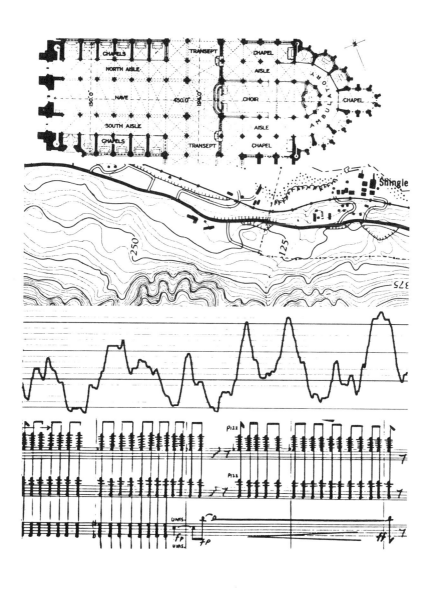

EIGHT

~~~~~~

Notation

Sound Images Notation is an attempt to render aural facts by visual signs. The value of notation for both the preservation and analysis of sound is therefore considerable, and it will be useful to spend a few pages discussing notational systems available to the modern researcher before letting our imaginations break forward again with the soundscape theme.

We have two descriptive techniques at our disposal already: we can talk about sounds or we can draw them. The first part of this book has been largely given over to talking about sounds because for a long time this was the principal means by which men attempted to study them, compare and classify them. Sounds resisted graphic representation for a long time and while we take it for granted that sounds may be described visually, the convention is recent, is by no means universal and, as I will show, is in many ways dangerous and inappropriate.

We have three graphic notational systems available:

1. that of acoustics, by which the mechanical properties of sounds may be exactly described on paper or a cathode-ray screen;
2. that of phonetics, by which human speech may be projected and analyzed;
3. musical notation, which permits the representation of certain sounds possessing "musical" features.

It is important to realize that the first two of these systems are descriptive —they describe sounds that have already occurred—while musical notation is generally prescriptive—it gives a recipe for sounds to be made.

The first attempt to give sounds graphic representation was the phonetic alphabet. Pictographs or hieroglyphs draw things or events, but phonetic speech draws the sounds of spoken words—that is the difference, and it represents not only a great advance in the versatility of writing but a radical departure in shifting the focus from the external world to the lips of the speaker.

Musical notation was the first systematic attempt to fix sounds other than those of speech, and its development took place gradually over a long period from the Middle Ages to the nineteenth century. From writing, music borrowed the convention of indicating time by movement from left to right. It introduced a new dimension, the vertical, by which frequency or pitch was indicated, high sounds being up and low sounds down. The matter is largely arbitrary, for while it is customary to point out that there are solid cosmological reasons for such a convention in that shrill sounds like those of birds come from the air while deep sounds come from the earth, thunder does not speak with a soprano voice, the mouse is not a baritone or the rattlesnake a timpanist.

The theoretical vocabulary of music has borrowed many indications from the visual arts and the world of spatial appearances: *high, low, ascending, descending* (all referring to pitch); *horizontal, position, interval* and *inversion* (referring to melody); *vertical, open, closed, thick* and *thin* (referring to harmony); and *contrary* and *oblique* (referring to *counterpoint*—which is itself a visual term). Musical dynamics also preserve traces of their visual origin—for instance, spreading lines indicate *crescendo,* or growing louder, while converging lines indicate the opposite, *diminuendo.* In "The Graphics of Musical Thought" (*Sound Sculpture,* Vancouver, 1975) I discussed how the habit of writing music down on paper made possible numerous devices and forms in Western music which were taken over from the visual arts and architecture.

The dilemma of conventional musical notation today lies in the fact that it is no longer adequate to cope with the meshing of the worlds of musical expression and the acoustic environment, which I have already identified as probably the most significant music-fact of our century, or at least one which must be grasped by the acoustic designer of future soundscapes.

The descriptive notations of acoustics and phonetics are much more recent in development and may really be said to have originated in the twentieth century. For sounds to be given exact physical description in space, a technology had to be worked out by which basic parameters could be recognized and measured in exact, quantitative scales. These parameters were time, frequency and amplitude or intensity. The fact that these three parameters have been identified as in some sense basic should not lead us to believe that this is the *only conceivable* method by which a total description of the behavior of sound is possible. It is also an artificial convention and, like musical notation, suggests a disposition toward 3-D thinking. At any event the three chosen parameters should never be regarded as isolable or independent functions. At least as far as our perception of them is concerned, they are in constant interaction. For instance, intensity can influence time perceptions (a loud note will sound longer than a soft one), frequency will affect intensity perceptions (a high note will sound louder than a low one of the same strength) and time will affect intensity (a note of the same strength will appear to grow weaker over time)—to give just a few examples of interaction. In introducing students to the properties of sound I have noticed frequent confusion between notions as elementary as frequency and intensity and have come to the conclusion that the standard acoustic diagram is not only ambiguous but for some people, at least, may not correspond at all with the natural instincts of aural perception. The problems between acoustics and psychoacoustics may never be clarified as long as the 3-D acoustic image continues to be regarded as an inviolably accurate model of a sound event.*

Problems of the 3-D Acoustic Image   Machines listen differently than men do. They have exceptionally wide hearing range, fine sensitivity, and no listening preferences whatsoever. They inform us about their listening abilities on paper printouts or cathode-ray displays. Various projections are possible, but with most only two dimensions of sound are given at one time. Thus intensity (or amplitude) is plotted against time, frequency against amplitude, or time against frequency.

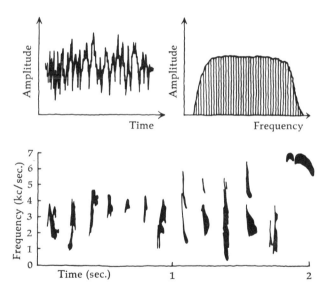

*The history of the technology of sound visualization would make a good subject for a thesis. Many of the people who worked on this problem came at the subject after work in visual studies. Typical was the case of Thomas Young, who invented the first practical means of projecting sound by means of a moving stylus connected to a tuning fork over a wax-coated revolving drum—an instrument called the phonautograph (1807). Young's previous work had been in the study of light (he was the first to measure astigmatism) and in the decipherment of Egyptian hieroglyphs.

The first two diagrams are for a typical broad-band noise, such as traffic; the third shows the relatively clear melodic pattern of bird-song. A real problem in these graphic projections is that only two dimensions of sound are presented simultaneously. The information is thus incomplete. While theoretically it ought to be possible to plot N + 1 dimensions in an N-dimensional space, in practice the placement of the three dimensions of sound on the two-dimensional space of paper results in formidable reading problems.

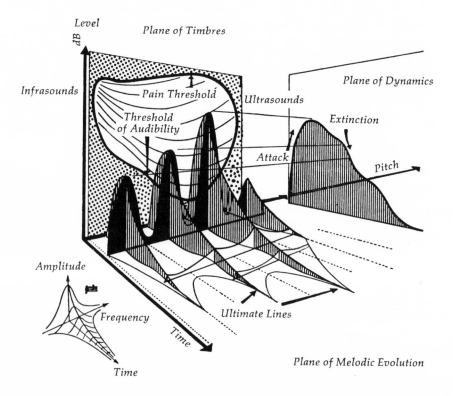

A three-dimensional representation of a simple sound object.

The sound spectrograph, an instrument developed at the Bell Telephone Laboratories in Princeton, New Jersey, incorporates all three dimensions of a sound, with intensity indicated by shading. Sound spectrograms thus render the sound image complete. The sound represented in the illustration is a Canadian Pacific train whistle and each shade of the contour spectrogram represents a five-decibel variation. But the relatively brief duration possible on the present-day spectrograph (a few seconds) renders it best fitted to represent single sound objects such as isolated bird calls, or juncture points such as the phonemic fluctuations of speech.

A sound spectrogram of a Canadian Pacific train whistle.

Furthermore, for easier reading, particularly strong formants or harmonic bands are often inked in, reducing the image again to a two-dimensional projection. (Such is the case in the preceding illustration representing bird-song.)

Acousticians Are the World's Best Sightreaders

I do not fault acoustic or phonetic machines for their inability to solve the problem of the simultaneous representation of the total sound image. That the two-dimensional image is sufficient for many kinds of investigation has, as I will be showing later on, a correspondence in our perceptual tendency to identify a limited number of significant features in any sound heard. If I have kept my enthusiasm for sound visualization under control up to this point, it is simply because I want the reader to remain alert to the fact that *all visual projections of sounds are arbitrary and fictitious.* This becomes emphatically explicit if we ask people to draw selected sounds when they are played to them on tape, to draw them in real time without forethought. In such exercises musicians or acoustical engineers often observe the conventions of left to right for time and up and down for frequency, while those without this kind of training react more independently. For them a sound may begin anywhere on the page. It may be coiled in a circle or be splattered about everywhere.

Hermann Helmholtz stood on the threshold between the aural and the visual study of sound. The most engaging feature of his monumental book *On the Sensations of Tone* (1877) is his great love of sound (he was a friend of musicians and himself a performer); but he could also write, concerning the study of vibration:

> To render the law of such motions more comprehensible to the eye than is possible by lengthy verbal descriptions, mathematicians and physicists are in the habit of applying a graphical method, which must be frequently employed in this work, and should therefore be well understood.

This strikes the pattern to be followed, and while the science of acoustics has advanced greatly since the nineteenth century, the listening abilities of average mortals have not shown corresponding improvement. In fact, they may have deteriorated in inverse proportion to the pictorialization of sound.

Today, many specialists engaged in sonic studies—acousticians, psychologists, audiologists, etc.—have *no* proficiency with sound in any dimension other than the visual. They merely read sound from sight. From my acquaintance with such specialists I am inclined to say that the first rule for getting into the sonics business has been to learn how to exchange an ear for an eye. Yet it is precisely these people who are placed in charge of planning the acoustic changes of the modern world.

A couple of years ago I was invited to speak at a symposium on transportation noise, organized by the U.S. Government. For several days acoustical engineers delivered papers on jet noise, fan noise, tire noise and so forth, illustrating their work with an ambitious array of slides and charts. Not a single sound was ever played as illustration. When I spoke, I began by reading back a catalogue of visual metaphors for sound from the researchers' own speeches: "You can see from the next slide that the sound has decreased in intensity"—that kind of thing. The shock of realization for those present was strong. Today acoustics is merely the science of sightreading.

I would not be dragging this point over so many paragraphs if I did not anticipate that we are on the threshold of a change. Such a change will be consistent with the theme announced in McLuhan's *The Gutenberg Galaxy:* "As our age translates itself back into the oral and auditory modes because of the electronic pressure of simultaneity, we become sharply aware of the uncritical acceptance of visual metaphors and models by many past centuries." If McLuhan is right, we may expect to move away from our dependence on visual representation of sound just as we are leaving print culture. It was print culture which, in McLuhan's view, moved the word away from its original association with sound and "treated it more as a 'thing' in space." Just as the notation of music is now being replaced by the record player, the tape recorder is pushing the physical study of acoustics into the human area of psychoacoustics. And for general field work in measuring the intensity of sound, the flat response of the sound level meter (the so-called C scale) has given way to the contoured A scale, weighted along the lines of the equal loudness contours of the human ear. Similarly, the numerous refinements in measuring aircraft noise in recent years have tended more and more to take human response into consideration. Thus EPNdB (Effective Perceived Noise in Decibels) takes into account the particular tone components of aircraft as well as the duration of the noise, while NNI (Noise and Number Index) uses the EPNdB as a basis and additionally calculates the number of aircraft per day (or night) as a key annoyance factor.

The acoustical engineer may not yet be a listener but he is at least adapting his instrumentation to hear more the way we do.*

Sound Objects, Sound Events and Soundscapes Since the Second World War, tape recorders have been abundant in broadcasting studios, and so it was that when the first psychoacoustic experiments were undertaken by the research group of the French radio in Paris in 1946, the director, Pierre Schaeffer, made good use of them. Although a mechanical engineer by training, Schaeffer never surrendered his ears for his eyes. This preoccupation with sound as sound is evident in his definition of the sound object (*l'objet sonore*), a term which he invented and defined as an acoustical "object for human perception and not as a mathematical or electroacoustical object for synthesis." We may call the sound object the smallest self-contained particle of a soundscape. Because it possesses a beginning, middle and end, it is analyzable in terms of its envelope. Envelope is a graphic term but the ear can be trained to hear its characteristics, which are defined as attack, body (or stationary state) and decay.

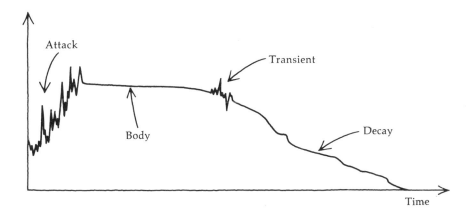

A few remarks about each of these components will not be out of place. The attack is the onset portion of the sound object. When a system is suddenly excited, an enrichment of the spectrum results, giving a rough or dissonant edge to the sound. Thus every attack of sound is accompanied by noise, and the more suddenly it appears, the more noise is present— a fact which is especially significant in electroacoustical systems with their brief switch-on times. When a sound develops more slowly, less of this

*On the other hand it seems necessary to point out that the proliferation of noise measurement systems, each claiming to be a refinement over the last, also tends to obscure the basic issue under a mask of jargon, designed largely to make it possible for engineers to stay in the noise pollution business without really solving it.

sudden spectral excitement is present and an even tone quality emerges. Many musical instruments have varying modes of attack, but some have a natural tendency to "speak" more quickly than others: compare the mandolin and the violin. The onset-transients of the attack may be only a few milliseconds long, but their importance in terms of characterizing the sound should never be underestimated. In fact, as Schaeffer and his colleagues demonstrated, when the attack portions of certain sounds are amputated, they may become wholly unintelligible or may be mistaken for others (a piano may then sound like a flute or a bassoon like a cello).

The middle portion of the sound object used to be called the stationary or steady-state portion, but it is better to call it the body, because nothing about sound is ever really stationary. Nevertheless, there may appear to be a period in the midlife of a sound when to the naked ear the sound seems unprogressive and stationary. Some sounds, such as bells, gongs, pianos and percussion instruments, have no apparent body, consisting exclusively of attack and decay. Other sounds, such as that of the air-conditioner, remain exclusively in the intermediate or stationary state. They do not die. This is an artificial condition, initiated, as I have already said, by the factories of the nineteenth century and extended by the Electric Revolution into all corners of modern living.

The bio-acoustic analogies I have just introduced are not merely personal ramblings, for the relationship between the two disciplines is made explicit in the term *decay*. The energy of a sound weakens; it withers and dies. There are rapid decays and there are infinitely slow decays.

The decay is usually combined with some sensation of reverberation. W. C. Sabine, the acoustician, has defined reverberation time technically. It is the time that elapses from the instant a sound source is switched off until its energy decays to one millionth of its original strength (a drop of 60 decibels). As far as the ear is concerned, it is the time it takes for a sound to melt and be lost in the ambient noise. Echo differs from reverberation in that it is a repetition or partial repetition of a sound, due to reflection off a distant surface. Reverberation is also reflected sound but no separate repetitions of the original are distinguishable.

Although the sound object may thus be subdivided for purposes of ear training, it must nevertheless always be considered integrally. Schaeffer: "A composed structure (such as we perceive it) cannot be deduced from separate perceptions of its component objects." But Schaeffer deliberately excludes all considerations of the sound object in any but physical and psychophysical terms. He does not want to confuse the study of sounds by considering their semantic or referential aspects. That a bell sound comes from a bell does not interest him. To him it is a phenomenological sound formation only. "The sound object must not be confused with the sounding body by which it is produced," for one sounding body "may supply a great variety of objects whose disparity cannot be reconciled by their common origin."

The limitations of such a clinical approach for soundscape studies will be obvious, and though soundscape researchers will want to be familiar with such work, we will be equally concerned with the referential aspects of sounds and also with their interaction in field contexts. When we focus on individual sounds in order to consider their associative meanings as signals, symbols, keynotes or soundmarks, I propose to call them *sound events,* to avoid confusion with *sound objects,* which are laboratory specimens. This is in line with the dictionary definition of *event* as "something that occurs in a certain place during a particular interval of time"—in other words, a context is implied. Thus the same sound, say a church bell, could be considered as a sound object if recorded and analyzed in the laboratory, or as a sound event if identified and studied in the community.

The soundscape is a field of interactions, even when particularized into its component sound events. To determine the way sounds affect and change one another (and us) in field situations is immeasurably more difficult a task than to chop up individual sounds in a laboratory, but this is the important and novel theme now lying before the soundscape researcher.

Aerial Sonography   The question is, which types of notation will be most helpful in these pursuits? At present there can be no grand solutions to this problem, for research is only beginning. It would be useful to have a notation or notations which could be read and comprehended immediately by professionals in many fields, particularly those on whom soundscape studies impinge most closely: i.e., architects, urbanologists, sociologists and psychologists as well as musicians and acousticians.

The best way to appreciate a field situation is to get above it. The medieval cartographer did this by climbing the highest hill, and the Mannerist painters of the Renaissance expanded the vistas of their paintings by doing the same thing. Surely one of the greatest inventions of man was the aerial projection in cartography, which represented a far bolder leap of the imagination than the fumbling exercises that eventually resulted in the actuality of flight.

One example of aerial projection applied to sound intensity is the isobel contour map. The isobel map derives from the contour maps of geographers and meteorologists, and consists of hundreds or thousands of readings on a sound level meter averaged out to produce bars of equal intensity, projected as if the observer were above the field of study. On such a map the quietest and noisiest sections of a territory can be immediately identified.*

Another type of aerial projection is the events map, which measures the distribution and recurrence of sounds. By means of events maps, comparisons can be made between two locations (say two blocks in different

---

*See Appendix I for isobel contour and sound event maps.

parts of a city) and the more persistent or characteristic sounds would be conspicuously revealed. Material for the events map would have to be limited to a specific period of time and would be gathered by walking over or around the selected location. (In the case of a city block this might be a single excursion around the block.)

Another example of aerial sonography, which brings value judgment into play, is that used by Michael Southworth in his article "The Sonic Environment of Cities." Here, after walking about the given territory freely, numerous observers were asked to comment on the sounds they heard, and the results of their observations were gathered together for display. The resulting map for a section of the downtown area of Boston shows where the acoustic designer could profitably begin his work.

Such diagrams are hints only, but perhaps this is all one should expect of sound visualization—a few hints which the ear can then follow up in its own way. It is easier for the inexperienced to absorb the salient information from them than from other types of graphic presentation, and that is to their advantage. The temptation of bad habits is no doubt still implicit in them, and it is for this reason that I conclude this chapter with a warning that no silent projection of a soundscape can ever be adequate. The first rule must always be: if you can't hear it, be suspicious.

NINE

~~~~~~

Classification

Why classify? We classify information to discover similarities, contrasts and patterns. Like all techniques of analysis, this can only be justified if it leads to the improvement of perception, judgment and invention.

Consider the dictionary—words slashed from their contexts and arbitrarily arranged according to their attack sounds. Yet, when used properly, the dictionary can contribute to the improvement of the language and can even provide us with inchoate thoughts and aesthetic moments.

Any classification system or taxonomy is surrealistic; for surrealistic art also depends on bringing together incongruous or anachronistic facts, which nevertheless somehow snap together to illuminate new relationships. The first such artists were the encyclopedists, who brought together strange groups of animals, vegetables and ideas for surrealistic family portraits.

Sounds may be classified in several ways: according to their physical characteristics (acoustics) or the way in which they are perceived (psychoacoustics); according to their function and meaning (semiotics and semantics); or according to their emotional or affective qualities (aesthetics). While it has been customary to treat these classifications separately, there are obvious limitations to isolated studies. My colleague Barry Truax puts the problem this way:

> Disintegrating a total sound impression into its component parameters appears to be a skill that must be learned; and while it is probably one that is necessary for acoustic design, a soundscape cannot be understood merely by a catalogue of such parameters, even if that were possible, but only through the representations formed mentally that function as a basis for memory, comparison, grouping, variation and intelligibility.

In this chapter I introduce some cataloguing systems for sounds—those systems which seem to be useful for dealing with various aspects of the soundscape—and the chapter will end with a discussion of the chief problems remaining to be solved. These have principally to do with the integration of classification systems. If soundscape study is to develop as an interdiscipline, it will have to discover the missing interfaces and unite hitherto isolated studies in a bold new synergy. This task will not be accomplished by any one individual or group. It will only be accomplished by a new generation of artist-scientists trained in acoustic ecology and acoustic design.

Classification According to Physical Characteristics

Let us consider first a physical classification for sound objects. Pierre Schaeffer has spent much effort in devising such a system. Schaeffer's concern is not really with acoustics but rather with psychoacoustics. He has tried to work out a paradigm by which it would be possible to classify all musical sound objects for the purpose of helping students to perceive their significant features clearly. He calls this a "solfège des objets musicaux." In his book he presents the paradigm in a table covering four pages. There are nearly eighty blocks in the table and many are further subdivided in a dazzling performance of French complexity. It would be useless to reproduce this table without Schaeffer's several-hundred-page explanation and rationale. The paradigm, it should be stressed, only deals with single musical sound objects. To cope with compound or more extended sound sequences, either the chart would have to be extended or the sounds would have to be broken up.

The system may be useful for the detailed analysis of isolated sound objects, but I would like to suggest a modification of it which might help to render it more immediately useful for soundscape field work. The idea would be to have a card on which the salient information of a sound heard could be quickly notated to be compared with other sounds. In line with our desire to comprehend sounds as events as well as objects (p. 131), it would be useful first to give some general information on setting: the distance of the sound from the observer, its strength, whether it rises clearly out of the ambiance or is barely perceptible, whether the sound under consideration is semantically isolable or is part of a larger context or message, whether the general texture of the ambiance is similar or dissimilar, and whether environmental conditions produce reverberation, echo or other effects such as drift or displacement.*

A chart could then be produced consisting of the answers to these questions plus a general physical description of the sound itself. For this purpose we might use a two-dimensional approach. On the horizontal

---

*Drift (fading) or displacement (ambiguous point of origin) often result from atmospheric disturbances such as wind or rain.

plane we will preserve the three components of the sound object discussed in the last chapter: *attack, body* and *decay.* On the vertical plane we will determine the relative *duration, frequency* and *dynamics* of the sound, to which we will add observations on any momentary *internal fluctuations* (technically called transients) and two new features, derived from Pierre Schaeffer: *mass* and *grain.*

These last two need an explanation. Mass is related to frequency. While some sounds consist of clearly defined frequencies or pitches, others consist of inextricably entangled frequency clusters. Such may be the case with the broad-band noise of traffic, a flock of birds or the pounding of surf. Sometimes the sound will occupy a fairly narrow frequency band, sometimes it will be broad-band. The frequency spectrum of white noise will extend across the entire audio range (20 to 20,000 hertz), though it may also be filtered down to occupy a quite narrow range, at which point it may even appear "tuned," so that it could almost be hummed or whistled. The mass of a sound is where its bulk seems to lie. It is regarded as the predominant bandwidth of the sound. Indeed both mass and frequency are often present in environmental sounds and they may sometimes occupy quite independent positions in the spectrum, as would be the case with a sound consisting of a low throb and a high warble. As mass is composed of frequency clusters it can be indicated in the frequency block on our chart by drawing in its approximate shape.

Similarly, grain is a special type of internal fluctuation, one with a regular modulatory effect. It is accordingly contrasted with transients, which are isolated or irregular fluctuations. Grain gives texture; it roughens up the surface of the sound and its effects consist of tremolo (amplitude modulation) or vibrato (frequency modulation). The tempo of these modulations may vary from slow pulsing effects to rapid warbles of 16 to 20 impulses per second, at which time their grainy effect will be lost. Thus in grain, a tactile word, we again meet the convergence of the senses of touch and audition as individual impulses pass from their flicker state to smooth contours of pitched sound.

I have devised my own signs to indicate these various effects, as shown on the following chart.

SETTING

1. Estimated distance from observer:_____meters.
2. Estimated intensity of original sound:_____decibels.
3. Heard distinctly ( ), moderately distinctly ( ), or indistinctly ( ) over general ambiance.
4. Texture of ambiance: hi-fi ( ), lo-fi ( ), natural ( ), human ( ), technological ( ).
5. Isolated occurrence ( ), repeated ( ), or part of larger context or message ( ).
6. Environmental factors: no reverb. ( ), short reverb. ( ), long reverb. ( ), echo ( ), drift ( ), displacement ( ).

Physical Description	Attack	Body	Decay
Duration	⌐ sudden / ⟨ moderate / ⟨ slow / ⟨ multiple	non-existent / brief / moderate / long / continuous	⌐ rapid / ⟍ moderate / ⟍ slow / ⟍ multiple
Frequency/ Mass	very high / high / midrange / low / very low	——————————→	
Fluctuations/ Grain	steady-state / transient / multiple transients / rapid warble / medium pulsation / slow throb	——————————→	
Dynamics	ff very loud / f loud / mf moderately loud / mp moderately soft / p soft / pp very soft / f>p loud to soft / p<f soft to loud	——————————→	
	←———— Total Estimated Duration of Event ————→		

Description of a sound event.

The symbols employed in the chart are not intended as exact graphic analogues, but rather as a handy index of devices for students to use in notating the significant physical features of sounds quickly during ear training exercises. Comparison of the chief characteristics of different sounds might also reveal useful distinctive features for the study of sound symbolism. The chart is, of course, useful only for isolated sound events, but despite its limitations it will serve to throw many of the most conspicuous features of isolated sounds into relief, as we can show in some simple classifications.

BARK OF A DOG
1 20 meters
2 85 dB
3 Heard distinctly
4 Hi-fi, human
5 Repeated, irregular
6 Short reverb.

SONG OF A BIRD
1 10 meters
2 60 dB
3 Heard distinctly
4 Hi-fi, natural
5 Part of extended song
6 No reverb.

FOG HORN
1 1,000 meters
2 130 dB
3 Heard distinctly
4 Hi-fi, natural
5 Periodic repetition
6 Long reverb., displacement

CHURCH BELL
1 500 meters
2 95 dB
3 Moderately distinctly
4 Lo-fi, technological
5 Periodic repetition
6 Med. reverb., drift

TELEPHONE
1 3 meters
2 75 dB
3 Heard distinctly
4 Hi-fi, human
5 Repeated
6 No reverb.

MOTORCYCLE
(passing on highway)
1 100 meters-pass-100 meters
2 90 dB
3 Indistinctly-distinctly-indistinctly
4 Lo-fi, technological
5 Isolated
6 No reverb.

Classification According to Referential Aspects We have next to consider a framework which will allow us to study the functions and meanings of sounds. Most sounds of the environment are produced by known objects and one of the most useful ways of cataloguing them is according to their referential aspects. But the system used to organize such a vast number of designations will be arbitrary, for no sound has objective meaning, and the observer will have specific cultural attitudes toward the subject. Even a library cataloguing system is stylized and reflects the interests and reading habits of librarians and library users. The only framework inclusive enough to embrace all man's undertakings with equal objectivity is the garbage dump.

The framework I present here is that which we have been using for one of the sub-projects of the World Soundscape Project, an extended card catalogue of descriptions of sound from literary, anthropological and historical documents. The only way we have of gathering information about the soundscapes of the past is through earwitness accounts by those who were there. From the first part of the book, the reader will know that I derived a great deal of information from this catalogue, which now numbers several thousand cards. The catalogue headings are arbitrary and have been built up empirically, but they do at least accommodate all descriptions we have encountered to date.

BARK OF A DOG

SONG OF A BIRD

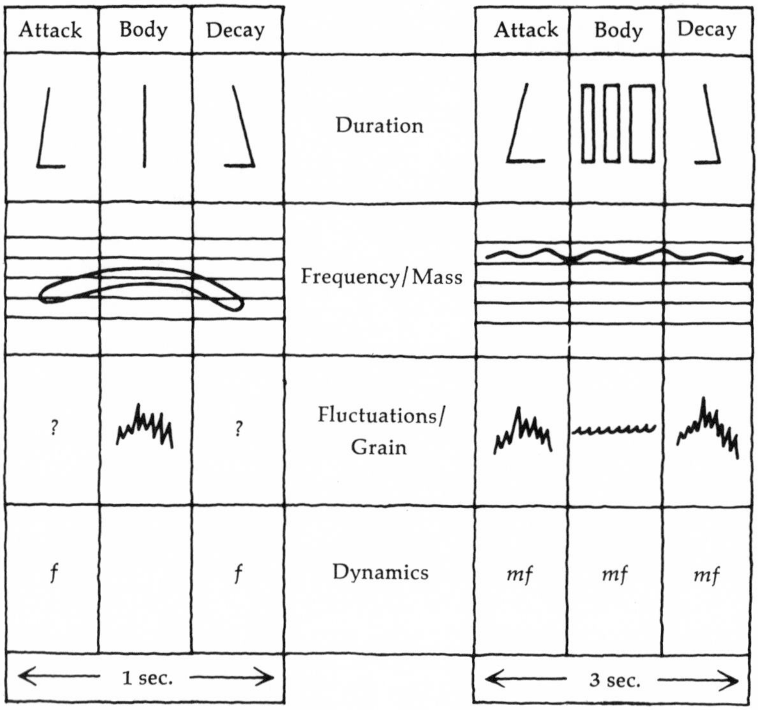

FOG HORN

CHURCH BELL

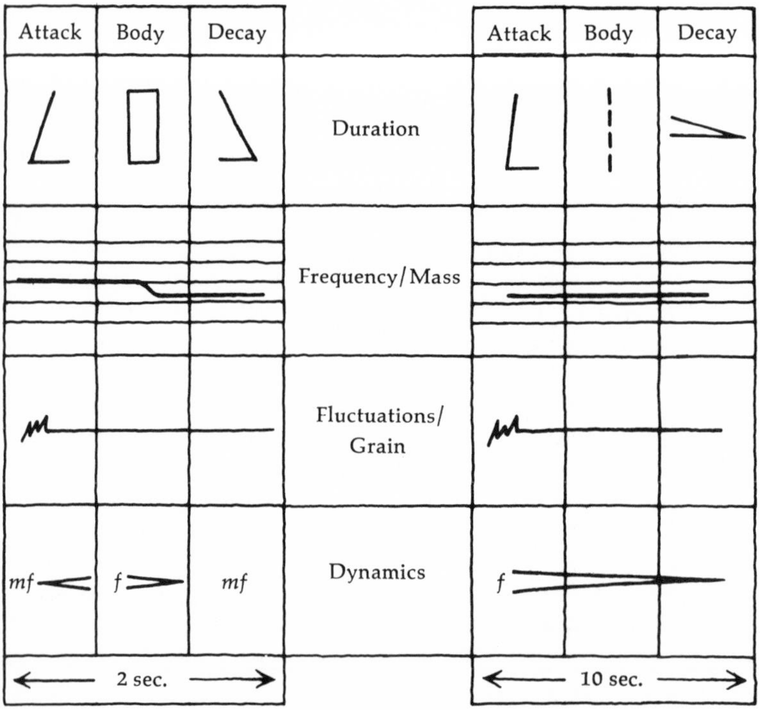

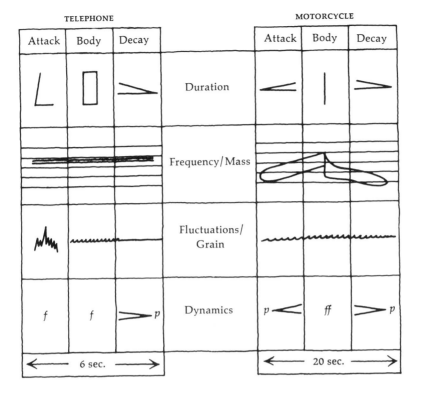

I. NATURAL SOUNDS

    A. SOUNDS OF CREATION

    B. SOUNDS OF APOCALYPSE

    C. SOUNDS OF WATER
        1. Oceans, Seas and Lakes
        2. Rivers and Brooks
        3. Rain
        4. Ice and Snow
        5. Steam
        6. Fountains. Etc.

    D. SOUNDS OF AIR
        1. Wind
        2. Storms and Hurricanes
        3. Breezes
        4. Thunder and Lightning. Etc.

    E. SOUNDS OF EARTH
        1. Earthquakes

2. Landslides and Avalanches
3. Mines
4. Caves and Tunnels
5. Rocks and Stones
6. Other Subterranean Vibrations
7. Trees
8. Other Vegetation

F. SOUNDS OF FIRE
1. Large Conflagrations
2. Volcanoes
3. Hearth and Camp Fires
4. Matches and Lighters
5. Candles
6. Gas Lamps
7. Oil Lamps
8. Torches
9. Festival or Ritual Fires

G. SOUNDS OF BIRDS
1. Sparrow
2. Pigeon
3. Killdeer
4. Hen
5. Owl
6. Lark. Etc.

H. SOUNDS OF ANIMALS
1. Horses
2. Cattle
3. Sheep
4. Dogs
5. Cats
6. Wolves
7. Gophers. Etc.

I. SOUNDS OF INSECTS
1. Flies
2. Mosquitoes
3. Bees
4. Crickets
5. Cicadas. Etc.

J. SOUNDS OF FISH AND SEA CREATURES
1. Whales
2. Porpoises
3. Turtles. Etc.

K. SOUNDS OF SEASONS
1. Spring

2. Summer
3. Fall
4. Winter

II. HUMAN SOUNDS

A. SOUNDS OF THE VOICE
1. Speaking
2. Calling
3. Whispering
4. Crying
5. Screaming
6. Singing
7. Humming
8. Laughing
9. Coughing
10. Grunting
11. Groaning. Etc.

B. SOUNDS OF THE BODY
1. Heartbeat
2. Breathing
3. Footsteps
4. Hands (Clapping, Scratching, etc.)
5. Eating
6. Drinking
7. Evacuating
8. Lovemaking
9. Nervous System
10. Dream Sounds. Etc.

C. SOUNDS OF CLOTHING
1. Clothing
2. Pipe
3. Jewelry. Etc.

III. SOUNDS AND SOCIETY

A. GENERAL DESCRIPTIONS OF RURAL SOUNDSCAPES
1. Britain and Europe
2. North America
3. Latin and South America
4. Middle East
5. Africa
6. Central Asia
7. Far East

B. TOWN SOUNDSCAPES
1. Britain and Europe. Etc.

C. CITY SOUNDSCAPES
1. Britain and Europe. Etc.

D. MARITIME SOUNDSCAPES
1. Ships
2. Boats
3. Ports
4. Shoreline. Etc.

E. DOMESTIC SOUNDSCAPES
1. Kitchen
2. Living Room and Hearth
3. Dining Room
4. Bedroom
5. Toilets
6. Doors
7. Windows and Shutters. Etc.

F. SOUNDS OF TRADES, PROFESSIONS AND LIVELIHOODS
1. Blacksmith
2. Miller
3. Carpenter
4. Tinsmith. Etc.

G. SOUNDS OF FACTORIES AND OFFICES
1. Shipyard
2. Sawmill
3. Bank
4. Newspaper

H. SOUNDS OF ENTERTAINMENTS
1. Sports Events
2. Radio and Television
3. Theater
4. Opera. Etc.

I. MUSIC
1. Musical Instruments
2. Street Music
3. House Music
4. Bands and Orchestras. Etc.

J. CEREMONIES AND FESTIVALS
1. Music
2. Fireworks
3. Parades. Etc.

K. PARKS AND GARDENS
1. Fountains
2. Concerts
3. Birds. Etc.

L. RELIGIOUS FESTIVALS
1. Ancient Greek
2. Byzantine
3. Roman Catholic
4. Tibetan. Etc.

IV. MECHANICAL SOUNDS

A. MACHINES (GENERAL DESCRIPTIONS)

B. INDUSTRIAL AND FACTORY EQUIPMENT (GENERAL DESCRIPTIONS)

C. TRANSPORTATION MACHINES (GENERAL DESCRIPTIONS)

D. WARFARE MACHINES (GENERAL DESCRIPTIONS)

E. TRAINS AND TROLLEYS
1. Steam Locomotives
2. Electric Locomotives
3. Diesel Locomotives
4. Shunting and Yard Sounds
5. Coach Sounds
6. Street Cars. Etc.

F. INTERNAL COMBUSTION ENGINES
1. Automobiles
2. Trucks
3. Motorcycles. Etc.

G. AIRCRAFT
1. Propeller Aircraft
2. Helicopters
3. Jets
4. Rockets. Etc.

H. CONSTRUCTION AND DEMOLITION EQUIPMENT
1. Compressors
2. Jackhammers
3. Drills
4. Bulldozers
5. Pile Drivers. Etc.

I. MECHANICAL TOOLS
1. Saws
2. Planes
3. Sanders. Etc.

J. VENTILATORS AND AIR-CONDITIONERS

K. INSTRUMENTS OF WAR AND DESTRUCTION

L. FARM MACHINERY
1. Threshing Machines
2. Binders

　　　　3.　Tractors
　　　　4.　Combines. Etc.

V. QUIET AND SILENCE

VI. SOUNDS AS INDICATORS

　　　A.　BELLS AND GONGS
　　　　1.　Church
　　　　2.　Clock
　　　　3.　Animal. Etc.

　　　B.　HORNS AND WHISTLES
　　　　1.　Traffic
　　　　2.　Boats
　　　　3.　Trains
　　　　4.　Factory. Etc.

　　　C.　SOUNDS OF TIME
　　　　1.　Clocks
　　　　2.　Watches
　　　　3.　Curfew
　　　　4.　Watchmen. Etc.

　　　D.　TELEPHONES

　　　E.　(OTHER) WARNING SYSTEMS

　　　F.　(OTHER) SIGNALS OF PLEASURE

　　　G.　INDICATORS OF FUTURE OCCURRENCES

Other categories in this system include Mythological Sounds, the Sounds of Utopias and the Psychogenic Sounds of Dreams and Hallucinations. We also have categories for the last sounds heard before sleep, the first sounds heard on waking and acoustic experiences that connect with the other senses (synaesthesia). The final section of the catalogue indicates whether the reporter showed a particular attitude to the sound(s) described. Was it considered as a signal, as noise, as painful, pleasurable, etc.?

As sounds may function in a variety of contexts, all descriptive cards, indexed in this system, are cross-referenced generously. Thus any given sound may appear in several places, allowing us the opportunity to regard it from several angles or to compare it with others of a similar set.

Playing with this index is a splendid listening exercise. Let me pull out a few cards dealing with the sounds of footsteps and you will hear what I mean. I have already mentioned how the felt boots of *Doctor Zhivago's* Russian winter seemed to "screech angrily" in the snow. Compare this with

- "the slap, slap of Gran's carpet slippers" (Emily Carr)
- "the clattering of the clogs" in Coketown (Dickens)
- "the loose tripping" feet of the Moroccans (Hans Ganz)

- "the violent clatter of . . . hobnailed wooden-soled shoes on the school flagstones" of a French provincial town (Alain-Fournier)
- "the flat, soft steps of the barefooted" (W. O. Mitchell)
- "the impish echoes of . . . footsteps" in the cloisters and quadrangles of Oxford (Thomas Hardy)
- or the way "the floor timbers boomed" under the strong rough feet of Beowulf.

By noting the date and place heard for every sound in the index, it is possible to measure historical changes in the world soundscape as well as social reactions to them. Then we can learn, for instance, that Virgil, Cicero and Lucretius did not like the sound of the saw, which was relatively new in their time (*c.* 70 B.C.), but that no one complained of factory noise until a hundred years after the outbreak of the Industrial Revolution (Dickens, Zola).

We can also note interesting proportional changes, for instance, between the number of descriptions of natural as against technological sounds. I am limiting the following observations to a period for which we have several hundred card samples. (It will be a long time before the index can be built up to a point where it may serve as a reliable indicator for all times and places.) Let us compare the nineteenth and twentieth centuries in Europe and America. We note that of all sound quotes from nineteeth-century Britain, 48 percent referred to natural sounds, while during the twentieth century, mentions of natural sounds had dropped to 28 percent. Among European authors the same decline is observed over the two centuries: 43 percent has dropped to 20 percent. Interestingly enough, this decline is not observed in North America (and our sample is very large here so that there can be little doubt about it); just over 50 percent of all quotes for both centuries refer to natural sounds. One might assume that North Americans are still closer to the natural environment, or at least have easier access to it than Europeans, for whom it definitely appears to be disappearing.

But the matter is not so simple. Our index does not show any corresponding increase in the perception of technological sounds throughout the same two centuries except for the period of the First World War, where the number increases sharply and then falls again. (The Second World War did not have a similar effect.) In fact, while the number of perceptions of technological sounds remains at the same level in Europe and Britain (about 35 percent of all observations), in America it actually declines!

But we also notice a decline in the number of times quiet and silence are evoked in literary descriptions. Of all descriptions in our file for the decades 1810–30, 19 percent mention quiet or silence; by 1870–90 mentions had dropped to 14 percent, and by 1940–60 to 9 percent. Thus it would appear that while writers are not consciously perceptive of the accumulation of technological sounds, at an unconscious level they are noticing the disappearance of quiet and silence. All this is perfectly consis-

tent with the keynote character of technological noise as I have been
describing it.

In going through the cards, I am struck by the negative way in which
silence is described by modern writers. There are few felicitous descriptions. Here are some of the modifiers employed by the most recent generation of writers: solemn, oppressive, deathlike, numb, weird, awful,
gloomy, brooding, eternal, painful, lonely, heavy, despairing, stark, sus-
penseful, aching, alarming. The silence evoked by these words is rarely
positive. It is not the silence of contentment or fulfillment. It is not the
silence toward which this book is modulating.

Classification According to Aesthetic Qualities Sorting
sounds according to their aesthetic qualities is probably the hardest of all
types of classification. Sounds affect individuals differently and a single
sound will often stimulate such a wide assortment of reactions that the
researcher can easily become confused or dispirited. As a result, study of
this problem has been thought too subjective to yield meaningful results.
Out in the real world, however, aesthetic decisions of great importance for
the changing soundscape are constantly being made, often arbitrarily. The
Moozak industry does not hesitate to make decisions about what kinds of
music the public is most likely to tolerate, nor did the aviation industry
consult the public before it entered on the development of the supersonic-
boom-producing aircraft. Acoustic engineers have also succeeded in introducing increasing amounts of white noise into modern buildings and have
invoked aesthetics in the process, by referring to the results as "acoustic
perfume."*

When such stupid decisions are being made almost daily, can the
systematic study of soundscape aesthetics continue to be ignored? If the
soundscape researcher is to assist in developing improved acoustic environments for the future, some kinds of tests will have to be developed for
the measurement of aesthetic reactions to sounds. At first they should be
kept as simple as possible.

Reduced to its simplest form, aesthetics is concerned with the contrast
between the beautiful and the ugly, so a good place to begin might be by
simply asking people to list their most favorite and least favorite sounds.
It would be good to know which sounds were especially pleasing or displeasing to people of different cultures; for such catalogues, which might
be called sound romances and sound phobias, would not only be of inesti-
mable value in a consideration of sound symbolism, but could obviously
give valuable directives for future soundscape design. Read in conjunction
with noise abatement legislation, sound phobias would also give a good
impression of whether a given by-law fairly reflected contemporary public
opinion concerning undesirable sounds.

*Acoustical engineering firms have also already taken over our term *soundscape* and speak
of "soundscaping an office" to refer to the same white-noise mesmerism.

One of the sub-projects of the World Soundscape Project has been to offer such a test in as many different countries as possible. We have tried to run the test in two parts. First, the subjects, who were mostly high school or university students, were simply asked to list the five sounds they liked best and the five they disliked most. Next we had them take a short soundwalk around their environment, and when they returned they were asked to repeat the assignment with specific reference to the sounds they had heard during the walk. I wish we had space to print the complete results to some of the tests, for they make a fascinating exercise in imagination and perception. Reducing them to the extent necessary for inclusion here can only be excused on the grounds that the general patterns produced support the hypothesis that different cultural groups have varying attitudes to environmental sounds.*

A few general observations are in order. First, climate and geography obviously influence likes and dislikes to some considerable extent. We note, for instance, that while in countries which touch the sea, ocean waves are well liked, in an inland country like Switzerland, the sounds of brooks and waterfalls are a much greater favorite. Where tropical storms may blow in suddenly from the sea, strong winds are disliked (New Zealand, Jamaica). It is also clear that reactions to nature are affected by the degree of proximity to the elements. As people move away from open-air living into city environments, their attitudes toward natural sounds become benign. Compare Canada, New Zealand and Jamaica. In the two former countries, the sounds of animals were scarcely ever found to be displeasing. But every one of the Jamaicans interviewed disliked one or more animals or birds—particularly at night. Hooting owls, croaking frogs, toads and lizards were mentioned frequently. Barking dogs and grunting pigs were also strong dislikes. The animal sound most universally liked was the purring of a cat.

While the Jamaicans had no attitude concerning machine sounds, these were strongly disliked in Canada, Switzerland and New Zealand. Jamaicans also approved of aircraft while the other nationalities did not. For all nations except Jamaica traffic noise was especially objectionable. There can be little doubt about this. From the present as well as similar tests we have run with smaller groups of other nationalities, it appears clear that technological sounds are strongly disliked in technologically advanced countries, while they may indeed be liked in parts of the world where they are more novel. I stress this finding because in attempts to confront the contemporary noise pollution problem I have frequently heard politicians and other opponents argue that we represent a minority, citing the case of the mechanic who enjoys a good motor or the pilot who enjoys listening to aircraft. But there can be no doubt that such attitudes form a small minority, at least among young people.

Among other striking cultural differences is the intense fondness of the Swiss for bells, while in other countries they are scarcely mentioned.

*See Appendix II for International Sound Preference Survey.

On the phobia side, the dentist's drill elicits some mention in all countries except Jamaica (where it is less familar?). But the sound of fingernails or chalk on slate is mentioned as a sound phobia in all countries, a matter to which we will return presently.

This test needs to be followed by others, more detailed. We need to find out with greater precision how and why different groups of people react differently to sounds. To what extent are the differences cultural? To what extent individual? To what extent are sounds perceived at all? The field is open for some intelligent testing on an international scale.

Sound Contexts   Throughout this chapter sound has been considered in separate compartments. Acoustics and psychoacoustics have been dissociated from semantics and aesthetics. It is traditional to divide the study of sound in this way. The physicist and engineer study acoustics; the psychologist and physiologist study psychoacoustics; the linguist and communications specialist study semantics, while to the poet and composer is left the domain of aesthetics.

ACOUSTICS	PSYCHOACOUSTICS	SEMANTICS	AESTHETICS
What sounds are	How they are perceived	What they mean	If they appeal
Physicist Engineer	Physiologist Psychologist	Linguist Communicator	Poet Composer

But this will not do. Too many misunderstandings and distortions lie along the edges separating these compartments. Interfaces are missing. Let us follow through a few specimen sounds to understand the nature of the problem. Consider first the sample pair of sounds in the following table.

SAMPLE SOUND	ACOUSTICS	PSYCHOACOUSTICS	SEMANTICS	AESTHETICS
Alarm bell	Sharp attack; steady-state with rapid amplitude modulation; narrow band noise on center frequency of 6,000 hertz; 85 decibels	Sudden arousal; continuous warble; high pitch; loud; decreasing interest; subject to auditory fatigue; sensitive pitch area	Alarm signal	Frightening, unpleasant, ugly
Flute music	Interrupted modulations of shifting frequency; near pure	Active patterned sound of shifting pitch;	Sonata by J. S. Bach; inducement to sit down and listen	Musical, pleasant, beautiful

SAMPLE SOUND	ACOUSTICS	PSYCHOACOUSTICS	SEMANTICS	AESTHETICS
Flute music (*continued*)	tones with some presence of even harmonics; varying between 500 and 2,000 hertz; 60 decibels	melodic contour; pure tones; highish register; moderately loud		

There are apparently no problems here. The two sounds are physically quite different and they accordingly have different meanings and draw forth different aesthetic responses. But even here the context can produce divergent effects. Thus, without altering the physical parameters of the sound, the meaning of the alarm bell could change if, for instance, it was only being tested. Knowing this, the listener would not be impelled to drop everything and run. Or, without changing the physical character of Bach's flute sonata, the aesthetic effect could be quite different if the listener did not like the flute or did not care for the music of J. S. Bach.

When we get discrepancies such as these, our reliance on automatic across-the-board equations falters, and we become aware of the fallacy that a given sound will invariably produce a given effect. Let us consider some more discrepancies. Two sounds may be identical but have different meanings and aesthetic effects:

SAMPLE SOUND	ACOUSTICS	SEMANTICS	AESTHETICS
Car horn	Steady-state, reiterative; predominant frequency of 512 hertz; 90 decibels	Get out of my way! I've just been married!	Annoying, unpleasant Festive, exciting

Or two sounds with quite different physical characteristics may have the same meaning and aesthetic effect:

SAMPLE SOUND	ACOUSTICS	SEMANTICS	AESTHETICS
I say, "Pierre, how are you?"	My crimpled baritone	Pierre is called.	Friendship
Margaret says, "Bonjour, Pierre."	Margaret's glorious contralto	Pierre is called.	Friendship

But supposing we are ringing up the Prime Minister of Canada, whose name is also Pierre. Margaret is his wife. I am not. Everything else remains the same, but the aesthetic effect is different:

SAMPLE SOUND	ACOUSTICS	SEMANTICS	AESTHETICS
Ditto	Ditto	Ditto	Annoyance
Ditto	Ditto	Ditto	Pleasure

Now consider the following pair of sounds:

SAMPLE SOUND	ACOUSTICS	PSYCHOACOUSTICS	SEMANTICS	AESTHETICS
Kettle boiling	Colored noise; narrow band (8,000+ hertz) steady-state; 60 decibels	High-pitched hissing sound	Tea is on.	Pleasing
Snake hissing	Colored noise; narrow band (7,500+ hertz); steady-state (occasionally intermittent); 55 decibels	High-pitched hissing sound	Snake preparing to attack	Frightening

Here two sounds with similar, but not identical, physical characteristics appear to be identical in perception, but nevertheless cause no confusion in meaning and accordingly have different aesthetic effects. Their contexts keep them clear. But when they are removed from their contexts in tape recordings, they may quickly lose their identities. Nor is the ear acute enough to be able to distinguish whatever differences may exist in their physical structure. Then the kettle may become the snake or either may become a green log on a fire.

It has always surprised me how even quite a common sound can be completely mistaken by listeners, dramatically affecting their attitudes toward it. For instance an electric coffee grinder was described as "hideous," "frightening," "menacing" by a group after listening to it on tape, though as soon as it was identified their attitudes immediately mollified.

There is one celebrated sound which seems to epitomize the interface dilemma which I have been describing: the sound of chalk or fingernails on slate. We have shown that it is an international sound phobia. Yet physical analysis fails to reveal why it should send cold shivers up the spine. It is not extraordinarily high or loud. It is not accompanied by any hurtful action. It does not even designate anything in particular. No single discipline then is capable of accounting for its remarkable effect. When sound enigmas like this are explained—and not until then—we will know that the missing interfaces are at last falling into place.

TEN

~~~~~~

Perception

It is not surprising, noting the visual bias of modern Western culture, that the psychology of aural perception has been comparatively neglected. Much of the work done has been concerned with binaural hearing and sound localization—which also has largely to do with space. Quite a lot of work has been done on masking (covering one sound by another) and some has been done on auditory fatigue (the effect of prolonged exposure to the same sound); but taken as a whole such researches leave us a long way from our goal, which would be *to determine in what significant ways individuals and societies of various historical eras listen differently.*

Thus it is inconceivable that a music or soundscape historian should get quite the same thrill out of the preparatory work the laboratories have provided as that which has stimulated art historians such as Rudolph Arnheim and E. H. Gombrich, whose work owes such a heavy debt to research in the psychology of visual perception. In the work of men like these it has begun to be possible to comprehend the history of vision, at least in the Western world. The soundscape historian can only speculate tentatively on the nature and causes of perceptual changes in listening habits and hope that psychologist friends may respond to the need for more experimental study.

Figure and Ground   It is indeed possible that some terms employed in visual perception may have equivalents in aural perception. At least they are probably worth careful examination. For instance, a phenomenon like irradiation—by which a brightly illuminated area seems to spread—does seem to have an analogy in that a loud sound will appear to be longer than a quiet one of equal duration. It is still not clear whether a term like *closure*—which refers to the perceptual tendency to complete an incomplete pattern by filling in gaps—can be applied to sound with anything like the confidence it has stimulated in visual pattern perception, though experiments in phonology show that for language at least there are striking parallels.

Throughout this book I have been using another notion borrowed from visual perception: figure versus ground. According to the gestalt psychologists, who introduced the distinction, figure is the focus of interest and ground is the setting or context. To this was later added a third term, *field,* meaning the place where the observation takes place. It was the phenomenological psychologists who pointed out that what is perceived as figure or ground is mostly determined by the field and the subject's relationship to the field.

The general relationship between these three terms and a set I have been employing in this book is now obvious: the figure corresponds to the signal or the soundmark, the ground to the ambient sounds around it— which may often be keynote sounds—and the field to the place where all the sounds occur, the soundscape.

In the visual figure-ground perception test, the figure and ground may be reversed but they cannot both be perceived simultaneously. For instance, looking into the clear water of a pond, one may perceive one's own reflection or the bottom of the pond, but not both at the same time. If we are to pursue the figure-ground issue in terms of aural perception, we will want to fix the points when an acoustic figure is dropped to become an unperceived ground or when a ground suddenly flips up as a figure—a sound event, a soundmark, a memorable or vital acoustic experience. History is full of such examples and this book is revealing a few of them.

Whether a sound is figure or ground has partly to do with acculturation (trained habits), partly with the individual's state of mind (mood, interest) and partly with the individual's relation to the field (native, outsider). It has nothing to do with the physical dimensions of the sound, for I have shown how even very loud sounds, such as those of the Industrial Revolution, remained quite inconspicuous until their social importance began to be questioned. On the other hand, even tiny sounds will be noticed as figures when they are novelties or are perceived by outsiders. Thus Lara notices the noise of the electric lights in Moscow as soon as Pasternak moves her in from the country (*Doctor Zhivago*) or I notice the scraping of the heavy metal chairs on the tile floors of the Paris cafés each time I visit that city as a tourist.

The terms *figure, ground* and *field* provide a framework for organizing experience. As useful as they may be, it would be injudicious to presume that they alone could lead to the goal announced at the beginning of the chapter, for they are themselves the product of one set of cultural and perceptual habits, one in which experience tends to be organized along perspective lines with foreground, background and distant horizon. How accurately they may apply to another society, remote from this one, is the big question we want answered.

Sonological Competence The psychologist studies the processes of perception; he does not attempt to improve them. But to run his tests he must assume some competence on the part of his subjects. As a teacher of music, my instinct tells me why so little has been accomplished to date. To report one's impressions of sound one must employ sound; any other method will be spurious. Just as we accused acousticians of playing sound false by turning it into pictures, so we accuse psychologists of playing it false by turning it into stories. This is the limitation of sound-association tests where listeners are asked to describe their impressions of taped sounds in free-association narratives. Whatever the purpose of such tests, it can hardly be to provide a description of perception. The only way to check perceptions is to devise routines by which listeners can reproduce exactly what they hear. This is why the ear training exercises of music are so useful. The dancing of the tongue in onomatopoeic mimesis is another way to check perceptions. As part of the ear cleaning program, I devised many exercises of this sort; for instance, imitate with your voice the sound of a shovel digging into sand, then into gravel, then clay, then snow. This exercise is partly memory work, partly vocal facility. Matching another person's voice, say in the repetition of a name, is another exercise designed to improve the competence of subjects for acoustic reportage.

In Chapter Two I noted how different languages have special onomatopoeic expressions for familiar animals, birds or insects. Aside from the phonetic limitations of language, the obvious differences in such words *must indicate something* about the manner in which the same sounds are heard variously by separate cultures—or is it that the animals and insects speak dialects?

Impression is only half of perception. The other half is expression. Uniting these is intelligence—accurate knowledge of perceptual observations. With impression we accommodate the information we receive from the environment.* Impression draws in and orders; expression moves out and designs. Together these activities, and perhaps some others about which we are as yet less certain, make up what Dr. Otto Laske has called "sonological competence." Laske points out that sonological competence does not result from the mere reception of sensory information. "If that were so, (psycho)acoustic knowledge would be sufficient for design, but it isn't. The difference between psychoacoustic knowledge and sonological competence is exactly the difference between a 'knowledge of, or about' and a 'knowledge-to-do,' i.e., between a knowledge of sound properties and a capability for designing." Laske insists that sonological competence

*Piaget calls these two complementary aspects of perception "accommodation" and "assimilation," but I prefer the outgoing suggestion of "expression."

applies to the most rudimentary level of perception, and as such it lies at the base of all deliberate attempts at soundscape design.

It is certainly possible that some societies possess better sonological competence than others. The evidence of this book makes it more than an assumption that this was the case when the ear was more important as an information gatherer, and the elaborate earwitness descriptions in works like the Bible and *The Thousand and One Nights* suggest that they were produced by societies in which sonological competence was highly developed. By comparison, the sonological competence of Western peoples today is weak. We have ignored our ears, hence the noise pollution problem. But in addition to our ears and voices we have today an instrument which can be used to assist in reclaiming the abilities of aural discrimination—I mean the tape recorder. With this device sounds can at last be suspended, dissected, intimately investigated. More than that, they can be synthesized and it is in this that the full potentiality of the tape recorder is revealed as an instrument uniting impression, imagination and expression. The tape recorder can synthesize sounds impossible for the voice. Take, for instance, an earthquake. The best description of one I have ever encountered is that of a radio sound effects technician.

Time after time I have heard this item portrayed by a sudden welter of earth-shattering sound and ear-splitting screams. This is way off the mark. The earthquake effect is done in four separate parts, with a few seconds pause between each. Start with a low, shuddering rumble, bring up the gain slowly, hold for a second or two, then drop it back almost to zero. Make the sound itself by shaking two rubber balls around in a cardboard box and recording the sound at double-speed or, if you are able to do so, recording at 15 ips and playing back at 3¾ ips. Having recorded the first part of the "quake" (or "prelude" as it is known), follow on with one or two isolated crockery-smashes and mix-in once more to the rumbling effect, louder this time.

Now bring in a sudden sliding, crashing sound, with a tearing metallic "ring" about it. This can be achieved by dropping a quantity of small stones on to the sloping lid of a cardboard box. The lid should be held about a foot above the table surface with a glass jam-jar (lying on its side) at the lower end of the slope. The sound sequence, thus, is that the stones strike the lid of the box, slide down its surface and strike against the side of the jam-jar before coming to rest on the table top. Record the sound at absolute maximum gain. Double-speeding may improve the item still further by both lengthening the sound and giving it a "heavier" quality. Lastly, fade in the rumbling noises once more, hold, then fade to zero.

Incidentally, a most uncanny yet effective impression of brooding silence can be obtained between the individual portions of activity by recording *very* faintly, the sound of distant voices alone. "Panic" noises such as screaming and shouting, if desired, are best recorded *behind* the third "falling-debris" section which may be superimposed over it.

I have often discovered in teaching that one of the best ways to press students into an examination of their perceptions is to set them similar exercises in sound synthesis using tape recorders. It is then that ignored or carelessly perceived features of a total sound complex become immediately conspicuous.

Music as the Key to Aural Perception   Any investigator of the world soundscape would benefit from a knowledge of the history of music. It provides us with a large repertoire of sounds—in fact, the largest repertoire of past sounds (not excluding those of speech and literature, which are less trustworthy owing to the vagaries of orthography and phonetic changes in language). The study of contrasting musical styles could help to indicate how, during different periods or different musical cultures, people actually listened differently. For the experience of music shows us that different features or parameters seem to characterize each epoch or school: thus Arab music is noted for rhythm and melody, while that of Western Europe—at least over the last three hundred fifty years— has emphasized harmony and dynamics. To have a good ear, to have musicianship in any culture, means then to have proficiency in selected areas, and the ear training exercises of any musical culture determine what they will be.

A cross-cultural study of the relationship between musical expression and aural perception has never been undertaken, but it should not too long be delayed. It would be of great value in answering questions like these: how does a society regard the relationship between frequency, time and intensity? between continuity and interruption? between impact and steady-state sounds? between foreground and background? signal and noise? or noise and silence—which is to say, dynamism and rest?

Perspective and Dynamics   I will give one example of the complementary development of musical expression and aural perception drawn from a single culture, and will try to show how it has developed into a concrete listening attitude. The dimension I want to consider is dynamics, which has an appropriate visual analogue in perspective. Perspective was introduced into European painting during the fifteenth century, and became the predominant style following the works of Masaccio and Uccello. There is only one ideal point from which a perspective painting may be viewed: viz., the point of view. Perspective fastens the viewer to a position directly before the window of the picture frame.

When Giovanni Gabrieli composed his *Sonata Pian' e Forte* (literally, to be sounded soft and loud), he introduced perspective thinking into Western music. Before this date we have no record of dynamic contrast in music, by which we must not infer that it did not exist, but may deduce that it had not become an articulated desideratum of performance. Gabrieli's *piano* and *forte* were the first steps toward the quantification of sound level, just as the foot and furlong had earlier quantified space. Just as objects are rank-ordered in perspective painting, depending on their distance from the viewer, so musical sounds are rank-ordered by means of their dynamic emphasis in the virtual space of the soundscape. It is an equally deliberate illusion which centuries of training turned into a habit. The classical Western composer places sounds in high definition before the eye of the ear.

Just as perspective focusing is unique to Western art, the organization of music along various dynamic planes is special to Western music. In fact, it is surprising to learn how absent dynamic nuancing is in many musical cultures. Von Békésy reports that, in his experiments on loudness discrimination,

> ... one of the subjects was a gypsy violinist. In the early part of the experiment his difference limens were enormous, far out of the range of the other subjects. His pitch limens, however, had about the usual values. After much probing it finally developed that he was paying little attention to loudness changes, and the reason for this was that in gypsy music only the pitch is considered an important variable and the loudness is kept relatively uniform. After this situation was understood by the subject, and deliberate training in loudness perception was carried out, his loudness limens fell to normal values.

The same thing was discovered by Catherine and Max Ellis in their work with Australian aborigines. When they were asked to play softer, they simply stopped playing.

By contrast, the exaggerated dynamic plane of Western music allows the composer metaphorically to move sound anywhere from the distant horizon to the immediate foreground. This implication of enormous space and infinite outreach achieved its most striking expressions in the works of Wagner and Debussy. But the important question now is: can we observe any equivalent to these advanced dynamic practices in Western perceptual habits of gathering the soundscape into perspective formations?

If the reader will re-read some of the quotations from the early parts of this book, he will discover an answer to this question. One further example will have to suffice here, though it is a compelling one, for it comes from a person who, in the course of his work, has had to give much thought to the sonic environment—the sound effects technician.

> Faced with a bewildering medley of sounds, the problem is to select those which will illustrate the scene and the accompanying commentary or dialogue to the best advantage. To this end, I recommend what is known as the "three-stage plan." In explanation it is apt to sound rather restrictive, but all it does, in fact, is to impose certain practical

limits on the number of effects to be included, collectively, in any one scene and to decide the degree of prominence that each shall enjoy.

The "three-stage plan" divides the whole sound-scene (called "scenic") into three main parts. These are: The "Immediate," the "Support," and the "Background." The chief thing to bear in mind is that the "Immediate" effect is to be *listened* to while the "Support" and the "Background" effects are merely to be heard. . . .

The "Support" effect refers to sounds taking place in the immediate vicinity which have a direct bearing on the subject in hand, leaving the "Background" effect to its normal job of setting the general scene.

Take, for example, the recording of a commentary at a fun-fair. The "Immediate" effect would be the commentator's voice. Directly behind this would come the "Support" effects of whichever item of fairground amusement he happened to be referring to, backed, to a slightly lesser degree, by the "Background" effect of music and crowd noises.

The three-stage plan of the radio technician corresponds precisely to the classical layout of the orchestral score with soloist, concertino group and tutti accompaniment. And it corresponds to the dynamic listening plane from foreground to horizon which makes focused listening possible. Furthermore, though the point must be made cautiously, the three-stage plan bears a recognizable resemblance to the figure/ground/field division of the (Western) psychologist.

Many other societies never developed the habit of perspective viewing. The study of Eskimo, Chinese and Byzantine art shows how differently space was perceived by these peoples. The Chinese spread objects out over the entire drawing surface, suggesting broad peripheral vision—the opposite of perspective focusing. More curious is the Byzantine convention of reversed perspective, by which objects were frequently enlarged as they receded in space. The Eskimos, as Edmund Carpenter has shown, would often continue a drawing over the edge of the drawing surface onto the back of the material, considering it part of the same surface. Carpenter writes:

> I know of no example of an Aivilik describing space primarily in visual terms. They don't regard space as static and therefore measurable; hence they have no formal units of spatial measurement, just as they have no uniform divisions of time. The carver is indifferent to the demands of the optical eye; he lets each piece fill its own space, create its own world without reference to background or anything external to it. . . . Like sound, each carving creates its own space, its own identity; it imposes its own assumptions.

Carpenter feels the Eskimo's space awareness is acoustic.

> Auditory space has no favoured focus. It's a sphere without fixed boundaries, space made by the thing itself, not space containing the

thing. It is not pictorial space, boxed-in, but dynamic, always in flux, creating its own dimensions moment by moment. It has no fixed boundaries; it is indifferent to background. The eye focuses, pin-points, abstracts, locating each object in physical space, against a background; the ear, however, favours sound from any direction.

If Carpenter is right, Eskimo culture provides an example of the reverse situation from the European Renaissance; with the Eskimos acoustic space has influenced and even dominated visual space.

Gestures and Textures   We have noted several times how focused listening with its implication of distance separating the listener from the sound event is disintegrating before the sound walls of the modern world. The modern lo-fi soundscape possesses no perspective; rather, sounds massage the listener with continual presence. As the population of sounds in the world increases, soloistic gestures are replaced by aggregate textures. Textures and crowds are correlatives. The daily sight of a swiftly moving crowd must have constituted an effect to which the senses had to adapt at first. Only after a new visual technique had been mastered did crowds cease to be confusing and the city-dweller learn to scan them leisurely for a chance display or an interesting figure. Many of Baudelaire's poems reveal this perceptual habit, which was presumably new in his time. "Amid the deafening traffic of the town"—so begins Baudelaire's sonnet "À une passante"—there arises by chance, out of a throng of pedestrians, a woman who snaps the poet's senses to attention by her beauty.

This has happened to us all. We are not looking for anything and we find it. We are not listening to anything but suddenly, out of the commotion, a sound jumps forward to become a figure. It would be inappropriate to say that this type of "unfocused" listening did not exist in the past, but it is possible to say that the circumstances which encourage it are more present in the textures of the post-industrial soundscape.

The present-day increase in statistical exercises and probability theorizing of all kinds is also a reflection of this crowding, nor is it surprising that, precisely at this juncture in history, statistics has entered music as a technique for composition. Iannis Xenakis describes his theory of composition as stochastic. "Stochastics," he explains, "studies and formulates the law of large numbers." More to the point, Xenakis has drawn his inspiration directly from the observation of the contemporary soundscape. He writes:

> But other paths also led to the same stochastic crossroads—first of all, natural events such as the collision of hail or rain with hard surfaces, or the song of cicadas in a summer field. These sonic events are made out of thousands of isolated sounds; this multitude of sounds, seen as a totality, is a new sonic event. This mass event is articulated and forms a plastic mold of time, which itself follows aleatory and sto-

chastic laws. If one then wishes to form a large mass of point-notes, such as string pizzicati, one must know these mathematical laws, which, in any case, are no more than a tight and concise expression of chains of logical reasoning. Everyone has observed the sonic phenomena of a political crowd of dozens or hundreds of thousands of people. The human river shouts a slogan in a uniform rhythm. Then another slogan springs from the head of the demonstration; it spreads towards the tail, replacing the first. A wave of transition thus passes from the head to the tail. The clamor fills the city, and the inhibiting force of voice and rhythm reaches a climax. It is an event of great power and beauty in its ferocity. Then the impact between the demonstrators and the enemy occurs. The perfect rhythm of the last slogan breaks up in a huge cluster of chaotic shouts, which also spreads to the tail. Imagine, in addition, the reports of dozens of machine guns and the whistle of bullets adding their punctuations to this total disorder. The crowd is then rapidly dispersed, and after sonic and visual hell follows a detonating calm, full of despair, dust, and death. The statistical laws of these events, separated from their political or moral context are the same as those of the cicadas or the rain. They are the laws of the passage from complete order to total disorder in a continuous or explosive manner. They are stochastic laws.

There are times when one sound is heard; there are times when many things are heard. *Gesture* is the name we can give to the unique event, the solo, the specific, the noticeable; *texture* is then the generalized aggregate, the mottled effect, the imprecise anarchy of conflicting actions.

A texture may be said to consist of countless inscrutable gestures. They are like the one-celled bacteria which are perceptible only in masses or cluster formations. Thus the sound events in a texture come to be considered statistically as they are in the countless number of sound level surveys being undertaken by so many modern cities in the modern world where noise pollution has broken out of control.

But for the soundscape researcher, the aggregate should never be confused with the singular, for they are not at all the same thing. The soundscape researcher must always remember Zeno's paradox: "If a bushel of corn turned out upon the floor makes a noise, each grain and each part of each grain must make a noise likewise, but, in fact, it is not so."

The aggregate sound of a texture is not merely a simple sum of a lot of individualistic sounds—it is *something different.* Why elaborate combinations of sound events do not become "sums" but "differences" is one of the most intriguing aural illusions.

In the broad-band texture there is also another aural illusion, for in such a sound other sounds may often be heard. I remember when Bruce Davis and I were working on the composition *Okeanos,* which combines the natural polynoise of the sea with electronic sounds and voices reciting maritime poetry. After many hours of working with the tapes of waves, we often heard in them other parts of the program, submerged as it were, rising at moments to the level of perception, then being carried off again to oblivion by the cascading waters.

Psychologists are aware of this type of aural illusion. In his fascinating little book *Soundmaking,* Peter Ostwald reports on the effects of playing a recording of a baby crying, masked nine decibels by white noise, to a group of patients in a mental hospital. The listeners heard the baby cry variously as

- "a voice shouting, a man's voice trying to be heard, an agitated sound"
- ". . . someone yelling and echoing"
- "a noisy factory with somebody hammering"
- "tremendous machinery, dynamos and . . . people shouting at each other"
- "a high sound, ayee, ayee, like a trumpet."

The polynoise of the sea resembles the white noise of the laboratory. Thus, no two waves are the same, and even the same wave, played over repeatedly on tape, will continue to yield up new secrets to the imagination at each listening. "You never go down to the same water twice," says Heraclitus.

Many other sounds also seem to have these miraculous powers. The wind, for instance, may even surpass the sea in mischievousness, and as witness we recall the contradictory voices of Typhoeus, the wind god of Hesiod's *Theogony,* quoted in Chapter One. In his *Treatise on Painting,* Leonardo da Vinci comments on ". . . the sound of bells, in whose strokes you may find every word which you can imagine." The same thing has been found to exist in words repeated over and over until they hypnotize the mind, at which point they may give rise to new word-sounds. Such is the function of a mantra. Perhaps the reasons why certain sounds produce aural illusions will never be satisfactorily explained. And perhaps it is just as well that they should not be, for an explanation would reduce their rich attraction as sound symbols.

ELEVEN

~~~~~~~~~~~~~~~~~~~~~~~~~~~~~~~~~~

Morphology

Morphology is the study of forms and structures. It is a nineteenth-century word, first used by the evolutionists in studying the development of biological forms; but by 1869 it was also being employed by the philologists to refer to patterns of inflection and word formation.

As I shall be using the term, I intend it to apply to the changing forms of sound across time or space. If typologies are systems for classifying sounds according to their various forms or functions, morphology allows us to gather together sounds with similar forms or functions in chronological or geographical sequence in order that variations or evolutionary changes might become clear. Thus morphology gives us the techniques for both depth-boring and cross-sectioning. In other words, we might use the morphological technique to study the evolution of, say, factory whistles —showing how the physical parameters of the sound were altered over time; or we might compare the factory whistle with alternatives employed in different societies for a similar purpose; that would also be a morphological study.* In a sense the whole first part of this book has been an essay in the general morphology of the soundscape, but for a true morphological investigation it is necessary to draw special groups of like sounds together in sharp relief.

Harold Innis, in *Empire and Communications*, stumbled on a truth which his Gutenberg bias allowed him only partly to express: "Media that emphasize time are those that are durable in character, such as parchment, clay, and stone. . . . Media that emphasize space are apt to be less durable

*I am aware that there is a certain similarity between these two types of studies and what structuralists call a paradigmatic series and a syntagmatic chain, but I think it best not to use expressions which sound like rusty iron on the tongue.

and light in character, such as papyrus and paper." He might better have substituted "less durable in character such as *sounds,*" for the true character of sound in shaping societies is in its spatial spread, as we will understand clearly when we come to study the acoustic profile as a delineator of the community; and the real paradox is that although sounds are pronounced in time, they are also erased by time. This is the difficulty when we approach the temporal axis of soundscape morphology. We have too few reliable sonic artifacts from the past. It is like visiting an instrument museum only to discover that all the instruments are broken or inoperative. Soundscape morphology—at least up to the invention of the tape recorder—will always be largely a matter of guesswork. But even though we lack a desirably large data base for a thorough morphological study, the technique can be outlined in a general way.*

From Wood to Plastic   The first thing to be considered is the material basis of different cultures and societies. Each geographic area of the earth has special materials in abundance which are used in the fabrication of dwellings, utensils and artifacts: wood, stone, bamboo or metals. And as these materials are chipped, scraped, sawed, hammered or broken they give off their own characteristic sounds. I have already noted that in Central Europe the original building material was wood; then as the land was cleared, it became stone; today it is the endless belt of raw concrete that unites house, street, city and nation together. By contrast, the North American West Coast is moving directly from the era of wood into gray modernity, without experiencing the "stone" age.

Now how does man deal with wood? In his *Georgics,* Virgil has recorded a significant flash-point in the technology of wood dressing:

> Then came the rigid strength of steel and
>   the shrill saw blade
> (For primitive man was wont to split his
>   wood with wedges).

The shrill saw blade was a relatively new sound in Virgil's day (*c.* 70 B.C.) and the parenthetic thought of the second line expresses nostalgia for the older method of dealing with the material. Virgil's contemporaries, Cicero and Lucretius, also disapproved of the sound quality of the saw. Cicero refers to the unpleasant noise of "*stridor serrae,*" and Lucretius records "the harsh grating of the strident saw." The next development in wood dressing is also recorded by the critical ear of a modern poet, Ezra Pound, where he significantly embeds it in his war-maddened Canto XVII (*c.* 1930):

*In the more restricted areas of specialized field study, the morphological approach can be applied more systematically. See in particular our study *Five Village Soundscapes,* Vancouver, 1976.

> And the first thing Dave lit on when he got there
> Was a buzz-saw,
> And he put it through an ebony log: whhssh, t ttt,
> Two days' work in three minutes.

I have already noted how the stonemasons' hammers on the Takht-e-Jamshid in Teheran reminded me of a similar transition in dealing with stone, as the discrete impact sounds of chipping gave way to the steady-state growl of the cement mixer. (But the metallic clip of the hammer was revived when metal nails replaced the wooden dowel; and the jackhammer chisel became indispensable as soon as it was more economical to knock apertures in the poured concrete structure than to plan them in the first place.)

A study of the introduction of metals would tell us much about the morphology of sounding materials. For instance, about 5000–4000 B.C. copper and tin were fused to produce the important new sound of bronze, which later was to find its most heroic voice in cannons and church bells. Bronze was the original metal of Europe, the Middle East and China (from before the Shang dynasty, 1523–1027 B.C.). India, on the other hand, produced a different alloy: brass, a fusion of copper and zinc. The difference in the tone may be tested even today in the elaborate platters and bells produced in that subcontinent.

When iron-smelting began, about 1000 B.C., it provided new sounds both in the making process and in the products. One of Charlemagne's biographers becomes truly delirious on the subject of iron in a ringing description from the ninth century, which connects this masculine metal with the improved art of warfare.

> Then came in sight that man of iron, Charlemagne, topped with his iron helm, his fists in iron gloves, his iron chest and his Platonic shoulders clad in an iron cuirass. An iron spear raised high against the sky he gripped in his left hand, while in his right he held his still unconquered sword. For greater ease of riding other men keep their thighs bare of armour; Charlemagne's were bound in plates of iron. As for his greaves, like those of all his army, they, too, were made of iron. His shield was all of iron. His horse gleamed iron-coloured and its very mettle was as if of iron. All those who rode before him, those who kept him company on either flank, those who followed after, wore the same armour, and their gear was as close a copy of his own as it is possible to imagine. Iron filled the fields and all the open spaces. The rays of the sun were thrown back by this battle-line of iron. This race of men harder than iron did homage to the very hardness of iron. The pallid face of the man in the condemned cell grew paler at the bright gleam of iron. "Oh! the iron! alas for the iron!" Such was the confused clamour of the citizens of Pavia. The strong walls shook at the touch of iron. The resolution of the young grew feeble before the iron of these older men.

Glass provides another distinctive range of sounds. Glass had been introduced to Europe by the twelfth century, and by 1448 (according to Aeneas Sylvius de' Piccolomini) half the houses of Vienna had glass in their windows. Glass provides the feminine counterpart to iron, and its dulcimer tones were to be heard in the touching of goblets, which provided the acoustic accent to complete the mixed-media experience of the wine-tasting ceremony; and in the soft shimmering tones of the glass harmonica, which romanticists like Jean Paul employed to evoke their *pays de chimères.* And when glass is broken, it tears at the heart like the sob of a woman.

Glass also had much to do with the disappearance of wood as a European keynote, for glass-making, along with metal-smelting, made it necessary to level enormous areas of forest.

In the twentieth century glass began to be replaced, first by celluloid, and then by plastic—the all-purpose modern material of peerless pudency, with a voice like a thud.

From Feet to Air Tires   The sounds of transportation could be lined up for morphological investigation, and considering the amount of time every human being spends each day involved with this activity, the resulting keynotes could be lifted into the foreground so that their effects on our lives could begin to be appreciated. I have already touched on the various sounds of footsteps, from the bare foot to the different sounds of shoes—wooden, leather, cleated. . . . The sounds of feet are often used for ostentatious display both in walking and dancing. One thinks of the little ankle bells which Persians and Arabian women used to wear and which Mohammed warned against, or of the great bunches of leaves which the Australian aborigine men tie about their knees before dancing.*

It is not in the heartbeat that the pulse of society is to be measured, but in the choreography of footsteps. The moderate movement of the Italian foot has given us the *andante* of music (*andare,* to walk). Contrast with this the nimble *courante* of the courtier or the rough and heavy *pesante* of the stooped and starving peasant. Yes, it is all here. To know the momentum of a society, measure the footsteps of its citizens. Are they purposeful? reckless? metallic? shuffling or clodhoppery? Sometimes foot-steps may form a protest against the prevailing tempi of a society. In this respect, North Americans, who live in probably the fastest-paced society of all time, have become some of the world's most sluggish pedestrians. In fact, the increased velocity of the car decreased the tempo of the Ameri-can footstep.

The discrete impact sounds of footsteps are united in the continuous line of the wheel. Imagine the first wheel—the lumbering sound it made over the uncleared ground. Then think of its transformation to the light spoked wheel, the wheel belted with metal stripping, the heavy iron wheel

---

*The natural rustling of Maori women's flaxen skirts (*piu piu*) produces a similar susurra-tion of great beauty.

of the cannon or the snorting wheel of the steam engine—halfway between impulse and flat line—for it was not until the invention of the internal combustion engine that rhythm was wholly disengaged from locomotion in the final speed-up to pitched noise. But even the air tire produces variations: the fizz of the wheels in rain or the heavier hum of snow tires and the clatter of studs.

From Horn to Telegraph   Communications systems have undergone striking acoustic metamorphoses. All acoustic communications systems have a common aim: to push man's voice farther afield. They share another aim also: to improve and elaborate the messages sent over those distances. One of the first acoustic devices to give man an extended voice was the horn. The first horns were aggressive, hideous-sounding instruments, used to frighten off demons and wild animals; but even here we note the instrument's benign character, representing the power of good over evil, a character which never deserted it, even when it began to be used as a signaling device in military campaigns. We know that the Greeks and Romans used various kinds of horns and trumpets in warfare, but we have no precise knowledge of how they were employed. The first dialoguing horn with which we are familiar is the alphorn; it was the first telephone in Europe.

But in sheer sophistication the alphorn is surpassed by the telegraph drums of Africa. Two drums are used for this purpose (high and low) and although sometimes different types of strokes are used, more frequently the code employed is strictly binary. It might be supposed that such a limitation would make it impossible to communicate complex messages, but such is by no means the case. Whenever ambiguity might exist, redundancy is introduced to make the message clear. For instance, if the signal for *moon* and *fowl* is identical, consisting of two strokes on the high-pitched drum (as is the case with the Lokele tribe of the Congo), the meaning is made clear by adding an explanatory phrase to each word.

> The moon looks down at the earth
> *songe li tange la manga*
> H H LH HL HL LL   L L

> The fowl, the little one which says *kiokio*
> *koko olongo la bokiokio*
> HH LHH L LHLHL

Possessing both contour and impulse, the talking drums of Africa, which can be heard up to sixty miles on a quiet evening, fuse melody and rhythm together in what is probably the most elegant signaling system ever devised. By comparison, Roland's mighty Oliphant was a piece of barbarism.

Melody was elevated over rhythm in European communications systems with the introduction of the festive embellishments of the baroque *cor de chasse* and the Thurn and Taxis postal signals, only to be flattened out again by the tickertalk of Morse's telegraph. It was about 1930, just as the last postal horns were fading out in Germany, that radio substituted the indoor concert for outdoor, just as it also introduced the commercial to synchronize with the disappearance of the street crier's movable market. (Yawping over long distance had, of course, already been possible via the telephone.)

The introduction and dismissal of music in communications systems (or at least the preferences shown for rhythm against melody or vice versa) is a subject that should interest the acoustic designer, and it can only be studied when the various systems are hooked up in sequence.

From Ratchet to Siren   Just as one can make interesting deductions about the most important social institutions in varying communities merely by noting the tallest buildings, studying changes in the most salient community signals could form an interesting theme for morphological research. For one thing, if the intensity of a historical set of community signals is measured, a pretty accurate idea of changes in the ambient noise level of the community can be obtained.*

Let us consider the way in which fires were signaled at various times in various communities. For this work to be truly revealing it should be restricted to a single depth study in one culture, or else take the form of a temporal cross-section comparing the devices of many contemporaneous cultures; but I have not got the facts at my fingers to do this, so I can only touch on the theme in a general way.

In Mozart's day (1756–91), Vienna was quiet enough that fire signals could be given by the shouts of a scout mounted atop St. Stefan's Cathedral. In early North America, fire halls also had tall watchtowers for scouting. As early as 1647 the governor of New Amsterdam appointed wardens to patrol the streets of Manhattan at night, armed with rattles or ratchets to sound the alarm. (An interesting posthumous example of the same device was the ratchet with which each civil defense warden was armed in wartime London and which was to be sounded in the event of a German gas attack. The sound generated by such instruments is surprisingly loud and we have measured one at 96 dBA at 3½ meters.)

On English fire vehicles, gongs were originally employed. The bell came into use early in the twentieth century with the advent of motor-driven appliances.

> The siren was introduced only after World War Two by some brigades but the bell continued to be the traditional audible warning for fire appliances in the British Fire Service. . . . However, during the 1960's, due to worsening traffic conditions and the increasing use of larger and diesel-engined commercial vehicles . . . a number of tests were carried

*And much more inexpensively than by going to the acoustical engineers. See Chapter Fourteen.

out using four different warning devices. . . . Following these tests it was decided to standardize on the use of a two-tone horn for fire appliances. This was subsequently adopted for other emergency service vehicles, i.e., police and ambulance, and its use is now restricted to vehicles of the emergency services.

In 1964 the familiar two-tone horn was adopted and the intensity fixed at not less than 88 dBA at a distance of 50 feet under calm conditions.

Two newspaper clippings in my file show that Canadian cities shifted from bell to siren at a considerably earlier date.

Clang! Clang! See the fire apparatus clashing, dashing by in a shower of sparks; firemen hastily donning their helmets and rubber coats, men whose hearts beat vigorous and warm in life, men whose prospects are filled with bright hopes and expectancy!

The wild plunge down the streets, the frantic speed of the horses, the drivers strapped to the seats, men clinging to the hose carts, ladder wagons and engines like flies!

That was an account of Vancouver's first horse-drawn fire engine of 1899. But the motorized fire engines that emerged from the hall after 1907 were no longer the same.

A long wolf howl, a sudden stopping of traffic, and a motor fire truck goes screaming down the street, leaving behind a clear track into which people and vehicles pour as the waters of the Red Sea followed on the wake of the men of Israel. Over the steering wheel the driver bends. At his side crouches a man who whirls the crank of the siren, sending ahead its shivering cry of fear.

In North America the revolving disc siren is employed on all emergency vehicles: fire-fighting equipment, ambulances and police cars. Europe, on the other hand, relies on the two-tone siren, variously tuned to an interval of a minor third (common in Sweden), a perfect fourth (common in Germany) or a major second (common in England).

Since the introduction of the disc siren in North America, the principal change has been in the volume of sound output. We have measured the siren on a 1912 vintage vehicle at 88 to 96 dBA at 3½ meters. By 1960 siren intensity had risen to 102 dBA at 5 meters. In recent years a new type of yelping siren has been introduced for emergency vehicles, measuring 114 dBA at the same distance. The United States is now manufacturing a yelping siren for police car use which measures 122 dBA at 3½ meters. With such hectoring devices the police are hardly becoming more lovable.

Conclusions on the Value of Morphological Studies

I hope the theme of morphological studies is suggestive enough that it will eventually inspire more systematic research. The tape recorder makes such work on the contemporary soundscape entirely feasible; and in connection with laboratory analysis, recorded sounds could be assembled in sequence and their physical changes could easily be analyzed.

Sometimes changes seem to progress in a fairly orderly manner; at other times they are interrupted suddenly by what I can only call mutations. The replacement of the bell by the siren is an example. In view of the heavy symbolism which accrues to well-established sounds, the acoustic designer ought to weigh the matter very carefully before substituting a radically new sound for a traditional one.

Right now in many countries foghorns are being automated and the character of their sound is being quite transformed. The haunting bass of the familiar diaphone and typhon is to give way to an electric horn, higher in pitch, shorter in carrying power. Fishermen in Canada say they don't like it and can't hear it, but the Ministry of Transport has begun to dismantle the old horns on both the Atlantic and Pacific coasts.

Sometimes a new technique only partially transforms a familiar sound, as is the case with the electric siren which, while maintaining the same contour as the old disc siren, truncates the arc of its glissando by instant switch-on and switch-off. The tempo of the old siren is increased about fourfold in the yelp mode of the new electric model, and the grainy effect of the original device has disappeared as a new sound signal is gradually shaped out of the old.

It is still too early to know whether there are any morphological rules of soundscape change, such as have been observed in language development. Of equal value, as research proceeds, will be the detection of what might be called matrix sounds. I am thinking here of sounds with unvarying physical characteristics which occur in different cultures or recur throughout history, always with the same general meaning. A knowledge of matrix sounds could be as useful to the acoustic designer as a knowledge of geometrical forms is to the visual designer. Such sounds would also carry a powerful symbolism.

TWELVE

~~~~~~

Symbolism

The sounds of the environment have referential meanings. For the sound-scape researcher they are not merely abstract acoustical events, but must be investigated as acoustic signs, signals and symbols. A sign is any representation of a physical reality (the note C in a musical score, the on or off switch on a radio, etc.). A sign does not sound but merely indicates. A signal is a sound with a specific meaning, and it often stimulates a direct response (telephone bell, siren, etc.). A symbol, however, has richer connotations.

"A word or an image is symbolic," writes C. G. Jung, "when it implies something more than its obvious and immediate meaning. It has a wider 'unconscious' aspect that is never precisely defined or fully explained." A sound event is symbolic when it stirs in us emotions or thoughts beyond its mechanical sensations or signaling function, when it has a numinosity or reverberation that rings through the deeper recesses of the psyche.

In his book *Psychological Types,* Jung speaks of certain types of "symbols, which can arise autochthonously in every corner of the earth and are none the less identical, just because they are fashioned out of the same world-wide human unconscious, whose contents are infinitely less variable than are races and individuals." To these "first form" symbols, Jung gave the name "archetypes." These are the inherited, primordial patterns of experience, reaching back to the beginning of time. They have no sensible extensions themselves, but may be given expression in dreams, works of art and fantasy.

In this chapter I am going to try to show how certain sounds possess strong symbolic character and how some of the most ancient may act to invoke archetypal symbols.

Return to the Sea   Of all sounds, water, the original life element, has the most splendid symbolism, and so we loop back to pick up the first theme of Chapter One. Rain, a stream, a fountain, a river, a waterfall, the sea, each makes its unique sound but all share a rich symbolism. They speak of cleansing, of purification, of refreshment and renewal.

The sea has always been one of man's primary symbols in literature, myth and art. It is symbolic of eternity: its ceaseless presence. It is symbolic of change: the tides; the ebb and flow of the waves. Heraclitus said, "You never go down to the same water twice." It illustrates the law of the conservation of energy: from the sea, water evaporates, becomes rain, then brooks and rivers, and finally is returned to the sea. It is symbolic of reincarnation: water never dies. Nor does water respect the law of gravity, for it flows downward and evaporates upward. When angry it symbolizes, in the words of W. H. Auden, "that state of barbaric vagueness and disorder out of which civilization has emerged and into which, unless saved by the effort of gods and men, it is always liable to relapse." Auden continues: "The sea is where the decisive events, the moments of eternal choice, of temptation, fall, and redemption occur."

"For thou hadst cast me into the deep, in the midst of the seas; and the floods compassed me about: all thy billows and thy waves passed over me." (Jonah 2:3.) To be saved from the clutches of corruption and chaos, as Jonah was, is always interpreted as a rebirth; for the miracle of water is that it is at once both the eternal destroyer and the grand deliverer. Jung remarks: "Water is the commonest symbol for the unconscious. . . . Psychologically, therefore, water means spirit that has become unconscious. . . . The descent into the depths always seems to precede the ascent."

The Greeks distinguished between Pontos, the mapped and navigable, and Okeanos, the infinite universe of water. Pontos corresponds to the closed world of Euclidean geometry, Okeanos to mystery and tempestuousness—for a storm on an unknown sea could swallow up a ship without warning or trace. The primal chaos of Okeanos is well served by the sound of a stormy sea. When the sea is worked into anger, it possesses equal energy across the entire audible spectrum; it is full-frequencied white noise. Yet the spectrum always seems to be changing; for a moment deep vibrations predominate, then high whistling effects, though neither is ever really absent, and all that changes is their relative intensity. The impression is one of immense and oppressive power expressed as a continuous flow of acoustic energy. In a storm at sea, the sound is not articulated into waves. It is only in a boat that wave motion becomes audible, for the bulkheads groan and shudder violently as the ship rolls and pitches. (By this means I once timed the waves at between 6 and 11 seconds in a gale on the Pacific.)

The sea symbolizes brute power; the land, safety and comfort. The tension between them is made audible in the crashing of the breakers. No sound unites continuity and discreteness so effectively within its signature. Thus, as we move back to the shoreline, power gives way to regular beating and, in a miraculous manner, the sea begins to suggest its opposite—the discrete side of its signature—rhythmic order. Rhythm replaces chaos as the sea becomes benign. Finally, the sea hangs over the horizon as an expiring murmur, blending with the gentler expressions of music. Here is how Thomas Mann, born on the Baltic, recalled it in *Tonio Kröger:*

> ... he played the violin—and made the tones, brought out as softly as ever he knew how, mingle with the plashing of the fountain that leaped and danced down there in the garden beneath the branches of the old walnut tree. The fountain, the old walnut tree, his fiddle, and away in the distance the North Sea, within sound of whose summer murmurings he spent his holidays—these were the things he loved, within these he enfolded his spirit, among these things his inner life took its course.

Modern man is moving away from the sea. Ocean travel has given way to air travel. The sea, which is down from everywhere, has come to be treated as a trough into which pollutants are dumped. Avoiding "the sea's green crying towers," modern man, landlocked and with untroubled heart, imagines the sea a sound romance. (Our Sound Preference Survey showed this clearly; see pages 147–148 and Appendix II.) He believes that the ebb and push of the waves on the summer beach exist merely to rhyme with relaxed breathing. But modern man is losing touch with the suprabiological rhythms that make the sea so notorious as a trembling presence in ancient art and ritual. Do all memories turn into romance? If so, the sea is the first example.

The Deviousness of the Wind

The Deviousness of the Wind   By comparison with the barbaric challenge of the sea, the wind is devious and equivocal. Without its tactile pressure on the face or body we cannot even tell from what direction it blows. The wind is therefore not to be trusted. "The wind bloweth where it listeth, and thou hearest the sound thereof, but canst not tell whence it cometh, and whither it goeth." (John 3:8.) Jung speaks of the wind as the breath of the spirit.

> Man's descent to the water is needed in order to evoke the miracle of its coming to life. But the breath of the spirit rushing over the dark water is uncanny, like everything whose cause we do not know—since it is not ourselves. It hints at an unseen presence, a numen to which neither human expectations nor the machinations of the will have given life. It lives of itself, and a shudder runs through the man who thought that "spirit" was merely what he believes, what he makes himself, what is said in books, or what people talk about. But when it happens spontaneously it is a spookish thing, and primitive fear

seizes the naïve mind. The elders of the Elgonyi tribe in Kenya gave me exactly the same description of the nocturnal god whom they call "maker of fear." "He comes to you," they said, "like a cold gust of wind, and you shudder, or he goes whistling round in the tall grass" —an African Pan who glides among the reeds in the haunted noontide hour, playing on his pipes and frightening the shepherds.

There is some etymological basis to what Jung writes. The Old German word for soul was *saiwalô,* which may be a cognate of the Greek αιολος, meaning "quick-moving, wily or shifty."

The illusory nature of the wind finds its instrument in the Aeolian harp, whose haunting and elusive tones were so affectionately regarded by the romanticists. Novalis wrote: "Nature is an aeolian harp, a musical instrument whose sounds come from the plucking of higher strings within us." But at times the wind seems to have a downright evil character. What are we to make of winds such as the Föhn in Germany and the Chinook in America, which have been cited as the cause of aberrational behavior and even death, usually by suicide? In an interesting unpublished paper, Dr. Philip Dickinson of the Institute of Sound and Vibration Research at the University of Southampton mentions the case of an elderly woman who tried to commit suicide.

> Her reason for the attempt: a low throbbing noise, that she alone seemed to be able to hear.... The local health department was unable to hear or record anything at all. It was then found that many other people also heard the noise, but were afraid to say so. Hence "expert" advice was brought in. A noise consultant visited the area with his wife, who was medically trained, and although he could hear nothing, recorded the "nothing" he could hear. On analysis of the noise a distinct peak was discovered in the 30–40 Hz range. Following newspaper accounts of these tests, reports came in from all over the country of severe noise disturbance from a low throbbing noise.... Many of these were investigated and in all cases a distinct peak of noise was discovered in the 30–40 Hz range. The noise was audible to the sufferers mainly at night, especially so on cold winter mornings in a slight breeze and in conditions of temperature inversion. Never on a hot summer day with no wind or with a very stiff breeze. Attempts to find the origin of the noise pinpointed power transmission lines in many of the areas. In some of these lines the wooden posts were vibrating so much that it was painful to place the ear against them. Not all the places had transmission lines, in others it seemed the noise was amplified by houses and possibly thin trees!

Dr. Dickinson has attributed these low-frequency vibrations to the wind. Uncontrolled low-frequency vibrations have been credited with causing brain tumors, a matter which Dr. Dickinson also raises in connection with his study.

Illusory, capricious and destructive, the wind is the natural sound man has traditionally mistrusted and feared the most. We recall that Typhoeus was a devious god because he spoke with so many tongues. The trickery of the wind has continued right down to modern times as anyone who has tried to make tape recordings outdoors well knows.

The Mandala and the Bell

Perhaps no artifact has been so widespread or has had such long-standing associations for man as the bell. Bells come in a vast array of sizes and have been put to an incredible diversity of uses. Most may be said to function in one of two distinct ways: either they act as gathering (centripetal) or scattering (centrifugal) forces. This can be seen from the following partial table.

PLACE OR TRIBE	TYPE OF BELL (OR GONG)	PURPOSE	FUNCTION
Rome	Bronze gong	To drive away ghosts	Centrifugal
Steiermark (Austria)	Parish bell	To drive away storms	Centrifugal
Eifel Mountains (Germany)	Small hand bell	To keep evil spirits from dying person	Centrifugal
Pueblo Indians (Arizona)	Small bells	To exorcize witches	Centrifugal
England (Middle Ages)	Hand bell carried by priest to sickbeds	To drive away witches	Centrifugal
Vancouver (1895)	Small bell on wagon carrying smallpox victims	To warn passers-by of possible infection	Centrifugal
Tonga and Fiji Islands	Bells	To summon worshipers	Centripetal
Athens	Hand bell played by priest of Proserpina	To call people to sacrifice	Centripetal
Japan	Small jingling bells of newspaper boys	To attract customers	Centripetal
Israel, Persia, Arabia	Ankle bells worn by women	To attract men	Centripetal

Not all bells can be categorized so easily according to function. In the Middle Ages in Europe, knights wore little bells attached to their armor and women wore them jingling from their girdles. Centripetal? But what do we say about the court jester, whose cap was adorned with the same little bells? And then there are the countless bells attached to animals all over the world in order to inform their owners of their whereabouts, or to identify the lead animal.

The bell is hung round the neck of the most willing horse of the pack, and from that moment he takes the lead. Till he moves on, it is almost impossible to force any of the others forward. If you keep back your horse for a mile or two when on the march, and then give him the rein, he dashes on in frantic eagerness to catch up to the rest. Get hold of the bell-horse when you want to start in the morning, and ring the bell and soon all the others in the pack gather round.

In the same account we learn how the bell of the lead horses signaled the approach of another pack train along the narrow trails of the Rocky Mountains. The jingling bells, attached to horses' harnesses, sounded a festive note which Edgar Allan Poe caught in a famous poem.

> How they tinkle, tinkle, tinkle,
>   In the icy air of night!
> While the stars that oversprinkle
> All the heavens seem to twinkle
> With a crystalline delight,
>   Keeping time, time, time,
>   In a sort of Runic rhyme,
> To the tintinnabulation that so musically wells
>   from the bells, bells, bells,
>   bells, bells, bells, bells,
> From the jingling and the tinkling of the bells!

Such bells were an adornment to the soundscape in many parts of the world, until the internal combustion engine eliminated them. In some places their demise was also assisted by regulations. A Saskatchewan by-law (No. 10, 1901) stated: "Horses and cows shall not carry bells within the limits of Prince Albert." And from Russia we recall how the eccentric Prince Nikolay Bolkonsky of Tolstoy's *War and Peace* had all the animal bells on his estate tied up and stuffed with paper.

The church bell originally maintained both a centripetal and centrifugal function, for it was designed both to frighten away evil spirits and also to attract the ear of God and the attention of the faithful. In ancient times church bells were accorded rich symbolism by numerous Christian commentators.

The bell denotes the preacher's mouth, according to the words of St. Paul: "I am become as sounding brass or a tinkling cymbal." The hardness of the metal signifies the fortitude of the preacher's mind according to the passage, "I have given thee a forehead more hard than their forehead." The clapper of iron which, by striking on both sides maketh the sound, doth denote the tongue of the preacher, the which with the adornment of learning doth cause both Testaments to resound. The striking of the bell denoteth the preacher ought first of all to strike at the vices in himself for correction, and then advance to blame those of others. The link by which the clapper is joined or bound to the bell is meditation; the hand that ties the clapper, denotes the moderation of the tongue. The wood of the frame upon which the bell hangeth doth signify the wood of our Lord's Cross. The iron that ties it to the wood denotes the charity of the preacher, who, being inseparably connected with the Cross exclaims: "Far be it from me to glory, except in the Cross of the Lord." The pegs by which the wooden frame is joined together are the oracles of the prophets. The hammer affixed to the frame by which the bell is struck signifieth the right mind of the preacher by which he himself holding fast to the Divine commands, doth by frequent striking inculcate the same to the ears of the faithful.

Here is another explanation of the bell, no less heartfelt but quite different, from a time nearer to our own:

The whole air seemed alive. It was as if the tongues of those great cold, hard metal things had become flesh and joy. They burst into being screaming with delight and the city vibrated. Some wordless thing they said touched something so deep inside you that they made tears come. Some of them were given in memory of dead people. That's a splendid living memorial, live voices speaking for the dead. If someone were to die and you were permitted either to see or hear them, I think it would be best to hear their voice.

While the contemporary church bell may remain important as a community signal or even a soundmark, its precise association with Christian symbolism has diminished or ceased; and it has accordingly experienced a weakening of its original purpose.

Bells and gongs differ in important respects which must now be explained. To a considerable extent these differences correspond to a fundamental difference between Eastern and Western cultures. A bell is a hollow cup-shaped body of cast metal, usually bronze. Chinese bells are struck on the outside, often with wooden mallets, but the European bell is struck by a metal clapper hung inside. In fact, the Europeans developed the clapper to great size, its weight sometimes reaching 1,500 pounds, as is the case with the great bell of Cologne Cathedral. Since it takes some time for the blow of the clapper to overcome the inertia of the metal, it embraces within its signature a sharp attack followed by a rounded orb of swelling sound. A gong is made of hammered, malleable metal, flat, or approximately so, and is usually struck with a soft mallet. As with Chinese bells, the sharp attack of the clapper is absent from the oriental gong. The sound of a gong is therefore more mellow, more diffused, though if the instrument is thin the metal will shiver, producing a rich transient distortion of full-frequency noise. The sounds of the two instruments may be approximately compared in the following graphs.

The sounds of the very words *bell* and *gong* suggest something of the

CHURCH BELL   TEMPLE GONG

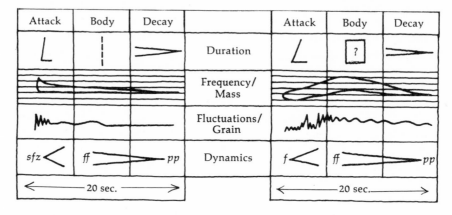

Attack	Body	Decay		Attack	Body	Decay
L		⟶	Duration	$\angle$	?	⟶
			Frequency/ Mass			
∿			Fluctuations/ Grain	∿∿		
sfz <	*ff* ⟶	⟶ *pp*	Dynamics	*f* <	*ff* ⟶	⟶ *pp*
← 20 sec. ⟶				← 20 sec. ⟶		

difference. The bell usually has a harder attack, *b,* and shorter decay, *ell;* the gong has a more subdued attack, *g,* and longer duration, *ong. Gong* is a Malay word of onomatopoeic origin, but *bell* derives from the Anglo-Saxon *bellam,* meaning to bellow. Kindred words are the Icelandic *belja* and the German *bellen,* meaning to bark. There is more aggression in the bell. If it did not actually provoke Western offensiveness, it is at least related to it, for Western history has witnessed a continual recasting of the same bronze from bells into cannons and back. During 1940, for instance, the Nazis confiscated 33,000 bells from churches in Germany and Eastern Europe for conversion into arms; and following the Second World War numerous churches and cathedrals (viz., St. Stefan's, Vienna) received back bells cast from cannons. The connection between these two seemingly antagonistic devices is emphatic and long-standing in European history.

Nevertheless, we are left with the fascinating fact that for a significant number of people, many of whom no longer find explicitly Christian associations in the church bell, the sound continues to evoke some deep and mysterious response in the psyche which finds its visual correspon-dence in the integrity of the circle or mandala. This is clear from tests we conducted in which subjects were asked to draw their impressions of sounds played to them on tape recordings. The sound of church bells frequently stimulated circular drawings. According to the psychologist C. G. Jung, the mandala symbolizes wholeness, completeness or perfection. One day perhaps we will be able to run a test similar to that with the church bell among oriental people, employing the gong as the test sound. With its less abrupt attack it would seem to be even more suitable to evoke the mandala image.*

*The fact that numerous people tested (in Canada) also produced rounded drawings in response to steady-state drones such as those produced by air-conditioners is perhaps ex-plained by my remarks to follow about the taming of natural sounds as man retreats into artificially controlled interior environments.

As the ambient noise of the modern city rises, the acoustic outreach of the church bell recedes. Drowned by merciless traffic, bells still possess a certain stammering grandeur, but the parish to which they now announce their messages has shriveled to a fraction of its once formidable size. By comparing earwitness accounts of the area over which church bells were once heard with the contemporary profile they create, this recession can be measured quite accurately. We have made such comparisons for the bells of Holy Rosary Cathedral in Vancouver and also for those in the village of Bissingen in Germany. By another method we verified the disappearance of church bells in the city of Stockholm. One evening in May, 1879, August Strindberg climbed the Mosebacke and wrote a detailed account of the sights and sounds of the city. Among the sounds, he gave particular attention to the city's seven church bells, describing their rings quite precisely. One evening nearly a hundred years later, a team of recordists from the World Soundscape Project climbed the Mosebacke and recorded the sounds of modern Stockholm from the same place. There were three church bells on the recording, one of them almost inaudible.

In many parts of Christendom, church bells are being eliminated altogether. While in the English city of Bath (population 100,000) there are 60 churches with 109 bells, our research has also revealed that of the 211 churches in Vancouver (population 1,000,000) 156 no longer have bells. Of those with bells, only 11 still ring them, though 20 have electric carillons or play recorded music. Significantly enough, the reason for silencing several of them was the complaint that they were contributing to noise pollution.

Horns and Sirens   As the bell declines, it is being replaced by the horn and the siren. One of the fundamental differences between the bell and the horn is that while the former radiates sound uniformly in all directions, the latter focuses or points it in a specific direction. The form of each instrument follows its function perfectly, for while the most effective acoustic shape for the horn is the logarithmic curve, which projects to infinity without ever turning back on itself, the shape of the bell resembles the normal or Gaussian distribution curve. Thus, the shape of the bell suggests community while that of the horn implies the outered projection of authority. It is the Oliphant of Roland, the bugle of the army or the whistle of the factory.

Long before the horn evolved into a family of musical instruments, it was employed as a kind of magical device, used by early men to frighten off evil demons. It was an aggressive, hideous-sounding instrument with supernatural capabilities. From the very beginnings it represented the power of good over evil. Thus, while it remained a commanding instrument of persuasion, it embodied also the blessings of victory and fulfillment. Contained within its broad masculine assertion of tone and curved flare is a feminine counterpart: the dark receding center of the bell.

We do not know who invented the horn, but the siren was invented

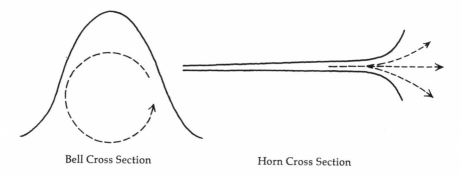

Bell Cross Section   Horn Cross Section

by Seebeck during the first half of the nineteenth century. Operating on the principle of the perforated disc, the siren, like the bell, radiates energy in all directions uniformly. Indeed it might have taken on the same numinosity had it not been put to quite different uses, for the siren broadcasts distress. It is a centrifugal sound, designed to scatter people in its path.

In Greek mythology the Sirens were nymphs who destroyed those who passed their island by means of their singing, at once piercing yet dulcet as honey. Circe warned Odysseus of the Sirens and so enabled him to elude their fatal song by plugging the ears of his men with wax and having himself bound to the mast of his ship. The Sirens thus signify mortal danger to man and this danger is broadcast by means of their singing. There is good evidence that the Greek word *siren* may be etymologically related to the words for wasp and bee. Modern man has reidentified the concept of danger with the wasp's song. There is an obvious similarity also between the glissando wail of the original siren and the human cry of pain or grief, diminished, however, since the introduction of the yelp siren with its sudden switch-on–switch-off technique.

Sirens and church bells belong to the same class of sounds: they are community signals. As such they must be loud enough to emerge clearly out of the ambient noise of the community. But while the church bell sets a protective spell on the community, the siren speaks of disharmony from within.

Symbolism in Transition   All acoustic symbolism, even that associated with archetypes, is slowly but steadily undergoing modification. Modern man has sought to escape both the wind and the sea by encapsulating himself in artificial environments. And just as he has sought to control the sea in the fountain, he has sought to tame the wind in the air-conditioner, for the ventilation systems of modern buildings are nothing more than techniques for getting the wind to blow in the right direction at the right force. Transformations such as these will undoubtedly change the symbolism of such archetypes. This is evidenced by the fact that while more ancient descriptions of the sea and the wind always stress their terrible aspect, in aesthetic preference tests today, both these natural elements appear as sound romances rather than sound phobias—except in places experiencing sudden, violent storms, such as Jamaica.

In his book *Technics and Civilization,* Lewis Mumford has pointed out how many inventions were first developed in the mines (elevator, escalator, railroad, artificial light, ventilation systems), and later were brought up to the surface to be put to wider use. Another chapter could now be added to this theme; for as modern man again sinks underground in his artificial and windowless environments, it is interesting to observe how many outdoor effects he contrives to take with him in their new synthetic guises. The list is long; it only begins with the fountain and the air-conditioner . . . and continues with plastic trees and stuffed flamingos . . . but no one yet knows where it may end.

Other natural sounds, rich in symbolism, have also undergone transformations. Thus thunder, the original *vox dei* and Sacred Noise, migrated first to the cathedral, then to the factory and the rock band. And bird-song, having been brought into thematic unity with the medieval garden, where its purpose was to orchestrate love, finally became transformed into the transistor radio, by which the contemporary Tristan and Isolde could groove to the "top fifty" in the backyards and parks of suburbia.

The sounds of machines took on a happy symbolism approximately two hundred years ago, when it was realized that they could release man from his immemorial bondage to the earth. Traditionally the machine symbolized two things: power and progress. Technology has given man unprecedented power in industry, transportation and war, power over nature and power over other men. Ever since the outbreak of the Industrial Revolution, Western Man has been infatuated with the machine's speed, efficiency and regularity, and with the extensions of personal and corporate power it afforded; and this enthusiasm for technological noise is now nascent in the rest of the world as well.

James Watt once stated that to most people, noise and power go hand in hand, though he did not like the idea. Today the hard-edged throb of motors can be heard around us continuously as the keynote of contemporary civilization, and whenever it has sprung into the foreground as figure, it has been glorified as the symbol of power and prosperity.

But there are ominous signs. We are just beginning (at least in the West) to realize that the fallout from unrestrained technological exploitation of the earth's resources is more frightening than first anticipated. As this idea gains more universal acceptance, we are discovering an unpleasant twist in the machine's symbolism, as demonstrated by the changes now taking place in international noise abatement legislation and practice.

For increasing numbers of people, the prevailing soundscape is that of city life. But the city itself is changing its tunes at an ever-increasing rate as the rage for new inventions increases. The effect is to push us into a

mood of nostalgia for disappearing and lost sounds. At the age of forty I
have many sound memories of the Canadian city which are no longer to
be heard (milk bottles, steam whistles, bicycle bells, horseshoes being
tossed against a metal spike). Everyone will have such a list. We listen back
à la recherche du temps perdu and notice how much has slipped away unperceived. Where? Where are the museums for disappearing sounds? Even the
most ordinary sounds will be affectionately remembered after they disappear. Their very ordinariness turns them into exceptional sound souvenirs.

Trapped in the only nostalgic moment of his entire life, the prisoner
of Albert Camus' novel *L'Étranger* vividly recalls the sounds of his native
city of Algiers.

> And, sitting in the darkness of my moving cell, I recognized, echoing
> in my tired brain, all the characteristic sounds of a town I'd loved, and
> of a certain hour of the day which I had always particularly enjoyed.
> The shouts of newspaper-boys in the already languid air, the last calls
> of birds in the public garden, the cries of sandwich-vendors, the
> screech of trams at the steep corners of the upper town, and that faint
> rustling overhead as darkness sifted down upon the harbour . . . one
> incident stands out . . . I heard the tin trumpet of an ice-cream vendor
> in the street, a small, shrill sound cutting across the flow of words.

Perhaps all sound memories turn into romances. And the more quickly
new sounds are hurled at us the more we are thrust back into the wells of
memory, attractively fictionalizing the sounds of the past, smoothing them
out into peaceful fantasies.

THIRTEEN

~~~~~~

Noise

When I first discussed the outline of this book with several publishers they were quite enthusiastic. "A book about noise pollution would be very timely," they said. I pointed out that I had already dealt with noise pollution in another publication* and that anyway there were already a large number of good books on the subject. When I went on to discuss the book I wanted to write, they grew uneasy. I insisted that the only realistic way to approach the noise pollution problem was to study the total soundscape as a prelude to comprehensive acoustic design. They assumed my interest was academic. I further suggested that multitudes of citizens (preferably children) needed to be exposed to ear cleaning exercises in order to improve the sonological competence of total societies, and went on to describe how, if such an aural culture could be achieved, the problem of noise pollution would disappear. They concluded I was a dreamer. Nevertheless, after years of association with the noise pollution issue, I have come to the realization that there are only two ways to solve it: the way I have just described, or a worldwide energy crisis. The largest noises in the world today are technological; thus the crack-up of technology would eliminate them.

Throughout this chapter I will draw extensively on a World Sound-scape Project study in which we examined by-laws and antinoise procedures from over two hundred communities around the world. The help of countless municipal officials who sent copious information in reply to our inquiries is gratefully acknowledged. The purpose of the survey was not to draft a model by-law (though we could probably do that) but to study the question of what constitutes noise in as many varied cultures as possi-

* *The Book of Noise*, Price Milburn Co., Wellington, New Zealand, 1973.

ble. Noises possess a great deal of symbolic character as sound phobias; and, in fact, the test of a good noise by-law would seem to be whether the most displeasing sounds of a given locale are effectively dealt with in the law. Before introducing this survey, however, some preliminary questions must be discussed.

The Evolving Definition of Noise

The increase in sound in the modern world has given rise to a change in the meaning of the word *noise*. Etymologically the word can be traced back to the Old French word *noyse* and to the eleventh-century Provençal words *noysa, nosa* or *nausa*, but its origin is uncertain. The suggestion that it may have originated in either of the Latin words *nausea* or *noxia* has been rejected. *Noise* has a variety of meanings and shadings of meaning, the most important of which are the following:

1. *Unwanted sound.* The Oxford English Dictionary contains references to *noise* as unwanted sound dating back as far as 1225.
2. *Unmusical sound.* The nineteenth-century physicist Hermann Helmholtz employed the expression *noise* to describe sound composed of nonperiodic vibrations (the rustling of leaves), by comparison with musical sounds, which consist of periodic vibrations. *Noise* is still used in this sense in expressions such as "white noise" or "Gaussian noise."
3. *Any loud sound.* In general usage today, *noise* often refers to particularly loud sounds. In this sense a noise abatement by-law prohibits certain loud sounds or establishes their permissible limits in decibels.
4. *Disturbance in any signaling system.* In electronics and engineering, *noise* refers to any disturbances which do not represent part of the signal, such as static on a telephone or snow on a television screen.

The matter is more complex than this. For instance, while the word *noise* was first used in English to imply "unwanted sound," it frequently took on a richer meaning and was sometimes used to imply "an agreeable or melodious sound." Chaucer uses it this way in his translation of the *Roman de la Rose.*

> Than doth the nyghtyngale hir myght
> To make noyse and syngen blythe. (ll. 78–79)
>
> Of whiche the water, in rennying,
> Gan make a noyse ful lykyng. (ll. 1415–16)

The King James version of the Bible also employs the word *noise* in a broad sense:

> Make a joyful noise unto the Lord, all ye lands. (Psalms 100:1)

While this wider connotation has disappeared from the English word today, it still exists with the French equivalent, *bruit;* for the Frenchman may still refer to the *bruit* of the birds or the *bruit* of the waves, as well as to the *bruit* of the traffic. One of the difficulties in dealing with noise internationally is that the word has slightly different nuancing in each language. I have also employed the word in a wider context in the expression "Sacred Noise" (see pages 51–52 and 114–115).

Of the four general definitions, probably the most satisfactory is still "unwanted sound." This makes *noise* a subjective term. One man's music may be another man's noise. But it holds out the possibility that in a given society there should be more agreement than disagreement as to which sounds constitute unwanted interruptions. "To disturb the public" then means to disturb a significant portion of the public, and it is in this manner that traditional legislation usually deals with noise problems. Such noise legislation may be called qualitative, inasmuch as it involves public opinion.

This contrasts with another type of legislation called quantitative. Such legislation sets decibel limits to specified undesirable sounds. If, for instance, a regulation states that the permissible level for an automobile is 85 decibels, an automobile producing 86 decibels is noisy while one producing 84 decibels is not noisy—or so the law would have us believe. The quantitative measurement of sound is thus tending to give noise a meaning as "loud sound." This is unfortunate because, as we know, not all irritating noises are necessarily loud, or at least loud enough to show up effectively on a sound level meter. Noise has come to be evaluated quantitatively owing to the risk of hearing loss, a matter about which enough is known for definite prevention criteria to have been established. It is, therefore, a subject that should be clearly understood.

The Hazards of Noise   Medical science has determined that sounds over 85 decibels, heard continuously over long periods of time, pose a serious threat to hearing. The resulting malady is often referred to as boilermaker's disease, because the earliest known victims were workers in factories where metal boilers were riveted together. Prolonged exposure to sound beyond this level may result, first, in temporary threshold shift (or TTS as it is sometimes called). TTS is an elevation of the threshold of hearing so that after being subjected to a very noisy experience, all sounds heard afterward seem fainter than usual. Normal hearing returns after a few hours or days. With further exposure, permanent cochlear damage may take place, resulting in permanent threshold shift (PTS). When this loss occurs in the inner ear, it is incurable.

Authorities concerned with industrial hygiene are now attempting to fix and enforce hearing risk criteria. In the U.S.A. a great step forward was taken when the Walsh-Healey Act of 1969 stipulated that no government contracts would be awarded to industries not respecting the established criteria. These were somewhat above the recommendations of the American Otological Society, and represented a compromise between the ideal and the immediately practicable. These recommendations parallel criteria already in effect in numerous European countries.

Permissible Noise Exposure
as Established by the Walsh-Healey Act (1969)

DURATION PER DAY (HOURS)	SOUND LEVEL (dBA)
8	90
6	92
4	95
3	97
2	100
1½	102
1	105
½	110
¼ or less	115

The threat of industrial hearing loss is now being resisted and is therefore not a concern of these pages. But PTS and TTS are by no means limited to these precincts. For instance, some researchers have found that exposure to levels as low as 70 dBA for 16 hours daily may be sufficient to cause a hearing loss. This is substantially lower than curbside traffic on a busy street. The term *sociocusis* has been devised to refer to non-industrial hearing loss, and a large number of examples could be given in illustration. For instance, it has been established by audiometric examination that persons operating power lawnmowers averaging 97 dBA suffer a temporary hearing loss after 45 minutes of exposure. We have already encountered a similar problem resulting from snowmobiles (Chapter Five) and amplified music (Chapter Seven). When Dr. George T. Singleton tested 3,000 public school children in Florida, he discovered a marked decrease in high-frequency hearing as the student progressed from the sixth to the twelfth grade, a period during which students had been exposed to rock bands, motorcycles and other "recreational" noises. Dr. Singleton and others found that the hearing ability of college freshmen who had attended rock concerts often deteriorated to that of sixty-five-year-olds.

Because sound is vibration it affects other parts of the body as well. Intense noise can cause headaches, nausea, sexual impotence, reduced vision, impaired cardiovascular, gastrointestinal and respiratory functions. But noises need not be intense to affect the physical state of humans during sleep. Russian researchers have found that "the level of thirty-five decibels can be considered as the threshold for optimum sleeping conditions ..." and that "when noise is at a level of fifty decibels ... there are fairly short intervals of deep sleep ... followed, on waking by a sense of fatigue accompanied by palpitations."

Everyone's hearing tends to degenerate a little with age. This happens very gradually and begins first in the high frequencies, which is the reason older people sometimes complain "everyone mumbles nowadays." This gradual loss of hearing acuity due to age is called presbycusis. It has always been assumed that presbycusis was a natural result of aging, like gray hair and wrinkles. This is now being challenged. A study on a tribe of Mabaan Africans in the Sudan showed very little hearing loss due to presbycusis. Africans at the age of sixty had as good or better hearing than the average North American at the age of twenty-five. Dr. Samuel Rosen, a New York otologist under whose supervision the study was made, attributed the superior hearing ability of the Africans to their noise-free environment. The loudest sounds the Mabaan heard were the sounds of their own voices singing and shouting at tribal dances.

How Fast Is the Ambient Noise Level Rising? In Chapter Five we saw how the noises of technology rubbed their way into both urban and rural life and how they were sanctioned as "progressive." By 1913 Luigi Russolo was able to point out that the new sensibility of man depended on his appetite for noise. Today, as the machines whirl in the hearts of our cities day and night, destroying, erecting, destroying, the significant battleground of the modern world has become the neighborhood Blitzkrieg. It is another reminder of the truth of Constantin Doxiadis's statement that for the first time in history we are less safe inside the city gates than outside them.

Precisely how fast the ambient noise level of the modern city is rising has been difficult to estimate. The figure of one decibel per year has frequently been given, but this seems excessively high when we remember that the decibel is a logarithmic term, so that a mere three decibels is approximately equal to a doubling of sound energy. In recent years, a great number of acoustical engineering surveys have been carried out in various cities in an attempt to determine the present noise level. It is an expensive proposition to do this properly, for thousands of readings must be taken by skilled workers using expensive equipment.

In order to point up the shortcomings of such surveys I will mention only one, though it is typical of them all. In 1971 Vancouver commissioned an extensive survey in which some ten thousand readings were taken over a grid stretching across the whole city region. The report concluded (in almost the only paragraph intelligible to the general public): "Traffic noise is the most significant noise source at all times. During the day hours local traffic noise was found to be responsible for 40% of all noise sources while distant traffic constituted some 13%. At night the corresponding values were 30% and 26%." Comparing their findings to those of similar surveys conducted elsewhere, the researchers concluded that the noise in Vancouver was some 6 to 11 dBA worse than that of some American cities in 1954. That would be an increase of about half a decibel per year. But this is not a particularly meaningful comparison, and the survey will only become useful if it is repeated in Vancouver in an identical manner at some later date. Given the rapid refinements of technical measuring, however, it is doubtful whether this will ever happen. Even then, without an integrating social survey to discover what the public thinks about the changing soundscape, the value of any engineering survey will remain under suspicion.

When a simple solution to a problem exists, an administrator will usually prefer a gummy one. I have already suggested that a simpler way to calculate the ambient noise increase would be to measure the sound signals of the community. The assumption would be that the level of ambient noise would rise in proportion to social signals, which must always remain above it. We did this for Vancouver by measuring the sound levels of different fire engine sirens, beginning with a 1912 La France device (88–96 dBA) and concluding with the newest 1974 siren (114 dBA), all measured at a distance of 3½ to 5 meters. This showed that the signals of emergency vehicles had risen some 20 to 25 decibels in sixty years, or nearly half a decibel per year on the average. The study complements and extends that of our acoustical engineering colleagues quite nicely, and extends our knowledge half a decade into the past. But, alas, few bellies were fed in the process.*

Public Reaction to the Rise in Ambient Noise If the ambient noise of the modern city is rising by something like half a decibel per year, what does the public think of it? One of the questions we asked municipal officials around the world was to list the noises receiving the most complaints from the public. The table below shows the total number of times each source was mentioned in each general category.

TYPE OF NOISE	NUMBER OF TIMES MENTIONED
Traffic (general)	115
Construction	61
Industry	40
Radios/Amplified music	29
Aircraft, etc.	28
Motorcycles/Motorbikes, etc.	23
Trucks	21
Animals	20
Bands/Discothèques	12
Parties	9
Power lawnmowers	7

*It may be pointed out that in a few isolated cities the noise level has actually been lowered by means of strict noise abatement procedures. Thus when Moscow prohibited the use of car horns in 1956, the result was a drop of 8 to 10 phons (Constantin Stramentov, "The Architecture of Silence," *The UNESCO Courier,* July, 1967, p. 11). The noise level in Göteborg (Sweden) has also been lowered by 7 dBA in recent years due to strict limits on new busses, compressors and garbage disposal trucks (Dr. B. Mollstedt, personal communication).

TYPE OF NOISE	NUMBER OF TIMES MENTIONED
Neighbors/People	7
Railroads	6
Shipyards	4
Snowplows	3
Snowmobiles	3
Church bells	2
Other	19

It will be more interesting to see how the complaints vary according to area. From numerous officials we obtained detailed reports on the number of complaints received for various categories of sound nuisance. Although the categories employed differ considerably, by reproducing the figures from six different cities on three continents some conspicuous differences can be observed.

London (England) 1969		*Chicago (U.S.A.) 1971*	
TYPE OF NOISE	NUMBER OF COMPLAINTS	TYPE OF NOISE	NUMBER OF COMPLAINTS
Traffic	492	Air-conditioners	190
Building sites	224	Construction	151
Telephones	200	Refuse trucks, etc.	142
Office machinery, etc.	180	Other trucks	125
Refuse vans	139	Factory noise	113
Street repairs	122	Musical instruments	109
Trucks (lorries)	109	Exhaust fans	97
Sirens	86	Loudspeakers	95
Ventilation machinery	69	Motorcycles	82
Voices	59	Automobiles	80
Motorcycles	52	Horns	77
Aircraft	42	Vibrations	55
Doors	34	Gas stations	34
Radios	10	Church bells	25
Railways	9	Trains	23
Factory machines	5	Miscellaneous	214
Miscellaneous	81		

Source: Report of the *Quiet City Campaign*, Port and City of London Health Committee, Guildhall, London, 1969.

Source: Department of Environmental Control, Chicago, Illinois.

Johannesburg (South Africa) 1972		*Vancouver (Canada) 1969*	
TYPE OF NOISE	NUMBER OF COMPLAINTS	TYPE OF NOISE	NUMBER OF COMPLAINTS
Animals and birds	322	Trucks	312
Amplifiers/Radios	37	Motorcycles	298
Construction	36	Amplified music/Radios	230
People	34	Horns and whistles	186
Machinery, etc.	29	Power saws	184
Home workshop	25	Power lawnmowers	175

Johannesburg (South Africa) 1972	
TYPE OF NOISE	NUMBER OF COMPLAINTS
Air-conditioning/ Refrigeration	19
Traffic	18
Musical instruments/Bands	15
Sirens	9
Milk deliveries	5
Mowers	2
Busses	1
Refuse collection	1
Vendors	1

Source: Noise Control Division, Medical
Health Department, City of Johannesburg.

Vancouver (Canada) 1969	
TYPE OF NOISE	NUMBER OF COMPLAINTS
Sirens	174
Animals	155
Construction	151
Automobiles	138
Jet aircraft	136
Small aircraft	130
Industrial	120
Hovercraft	120
Domestic	95
Foghorns	88
Trains	86
Children	86
Office noises	81

Source: *A Social Survey on Noise*, World
Soundscape Project, Simon Fraser University,
Burnaby, B.C., Canada.

Paris (France) 1972	
TYPE OF NOISE	NUMBER OF COMPLAINTS
Domestic and neighborhood noise	1,599
Construction and road works	1,090
Industrial and commercial noise	1,040
Restaurants and cabarets	553
Miscellaneous	90

Source: Bureau de Nuisances, Paris, France.

Munich (Germany) 1972	
TYPE OF NOISE	NUMBER OF COMPLAINTS
Noisy restaurants	391
Industrial noise	250
Construction	87
Traffic	29
Domestic noise	27
Aircraft noise	11
Miscellaneous	2

Source: Der Umweltschutzbeauftragte,
Landeshauptstadt München.

While these statistics have been differently organized, some quite intriguing variations emerge. Note for instance the difference between the chief complaint in London and Chicago; or that between the chief complaint in Johannesburg and Vancouver—two cities of approximately the same population and both in temperate climates. Note also the way the proximity of sea and forest have affected the types of complaint from Vancouver. Also of interest is the varying incidence of traffic noise complaints in the six cities. As the general world survey placed it indisputably at the top of the list of offensive sounds, some explanation is necessary.

Whether a person complains about a sound or decides to bear it may be partly conditioned by whether or not action can be expected as a result of the complaint. This at least was the experience in Chicago. In 1971 a new Chicago ordinance went into effect. It is one of the toughest and most comprehensive anywhere in the world. The immediate reaction to the new law was a dramatic increase in the number of complaints. In 1970 the city government received approximately 120 noise complaints. During the first six months of 1971 (before the new law went into effect) the number rose to approximately 220; but during the latter half of the year it soared to 1,300, and has been steadily climbing ever since.

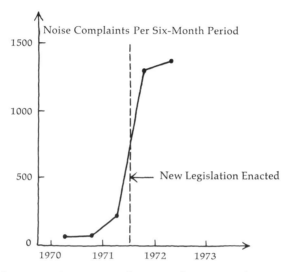

Increase of noise complaints in the city of Chicago.

Some Aspects of Noise Legislation

The only truly effective piece of noise legislation ever devised was in the form of divine punishment. In *The Epic of Gilgamesh* (*c.* 3000 B.C.) we read:

> In those days the world teemed, the people multiplied, the world bellowed like a wild bull, and the great god was aroused by the clamour. Enlil heard the clamour and he said to the gods in council, "The uproar of mankind is intolerable and sleep is no longer possible by reason of the babel." So the gods in their hearts were moved to let loose the deluge.

The first example of a by-law in the modern sense relating to noise was passed by Julius Caesar in his Senatus Consultum of 44 B.C. "Henceforward, no wheeled vehicles whatsover will be allowed within the precincts of the city, from sunrise until the hour before dusk. . . . Those which shall have entered during the night, and are still within the city at dawn, must halt and stand empty until the appointed hour." Due to crowding in the narrow streets, wagons were permitted to circulate only at night, which can hardly have assisted sleep. In his third *Satire,* Juvenal (A.D. 117) says: "It is absolutely impossible to sleep anywhere in the city. The perpetual traffic of wagons in the surrounding streets ... is sufficient to wake the dead."

By the thirteenth century, many towns in England had enacted laws restricting blacksmiths to special areas because of bothersome noise. In the same country, street music had been suppressed by two Acts of Parliament during the reign of Elizabeth I, and we have already mentioned Michael Bass's celebrated 1864 Bill against the same offense. Similar legislation was common to all countries of Europe. By selecting one city we can get a historical overview of the situation for Central Europe.

City of Bern (Switzerland) *

YEAR PASSED	BY-LAW
1628	Against singing and shouting in streets or houses on festival days
1661	Against shouting, crying or creating nuisances on Sunday
1695	Against the same
1743	For respect of the Sabbath
1763	Against disturbing noises at night
1763	Against noisy conduct at night and establishing regulations for night watchmen
1784	Against barking dogs
1788	Against noises in the vicinity of churches
1810	Against general noise nuisances
1878	Against noises near hospitals and the sick
1879	Against the playing of music after 10:30 p.m.
1886	Against the woodworking industry operating at night
1887	Against barking dogs
1906	For the preservation of quiet on Sundays
1911	Against noisy music, singing at Christmas and New Year's parties and against unnecessary cracking of whips at night
1913	Against unnecessary motor vehicle noise and blowing horns at night
1914	Against carpet-beating and noisy children
1915	Against beating carpets and mattresses
1918	Against carpet-beating and music-making
1923	For the preservation of quiet on Sundays
1927	Against noisy children
1933	Against commercial and domestic noises
1936	Against bells, horns and shouting of vendors
1939	Against excessive noises on holidays
1947	For the preservation of quiet on Sundays
1961	Against commercial and domestic noises
1967	For the preservation of quiet on Sundays

*I am grateful to Dr. G. Schmezer for providing information from the municipal archives.

Obviously it would be impractical to present a detailed analysis of all contemporary noise legislation here, so we will have to be content to touch on some of its aspects.

Many countries (for instance, Britain, France, Germany, Poland, Sweden, Turkey and Venezuela) have national legislation which may or may not be supplemented by municipal by-laws. In Britain, for instance, the Noise Abatement Act of 1960 and the Noise Insulation Regulations of 1973 are enforceable by every authority in England, Scotland and Wales. In other countries (such as Canada, Australia and parts of the U.S.A.) general legislation is drafted by the provinces or states for adoption or embellishment by municipalities. Elsewhere the matter may be left entirely up to the municipalities or arrangements between various levels of government may be worked out. Sorting out these peculiarities on an international scale is an exceedingly difficult if not impossible task, but I have tried to indicate some of the principal features of the various types of legislation in the following list. To show some general differences as clearly as possible, the communities have been arranged by continent.*

Civic Noise Abatement Legislation Around the World

○ No anti-noise legislation
⊙ Nuisance or other type of by-law with some reference(s) to noise
⊖ State or national health or environmental protection legislation, with some provision for noise
● Anti-noise by-law
□ Qualitative legislation
■ Quantitative legislation
⟶ Community considering or engaged in preparation of new anti-noise legislation
⊞⟶ Quantitative legislation planned

Africa

Banjul (Gambia) ⊙
Beira (Mozambique) ○
Bizerte (Tunisia) ●□
Blantyre (Malawi) ⊙
Bulawayo (Rhodesia) ⊙●□
Cape Town (South Africa) ●□
Durban (South Africa) ⊙●□

East London (South Africa) ○
Freetown (Sierra Leone) ○
Jadida (Morocco) ●□
Johannesburg (South Africa) ●□⊞⟶
Kimberley (South Africa) ○
Luanda (Angola) ○
Mombasa (Kenya) ⊙

*For the sake of clarity a number of smaller municipalities supplying information, notably in Australia and Canada, have not been included. For a detailed assessment of the situation in Canada the reader is referred to the World Soundscape Project document, *A Survey of Community Noise By-Laws in Canada (1972).* In the international survey we were unfortunately unsuccessful in securing sufficient accurate information from Communist countries. In the following table, communities with no legislation other than provision against noisy vehicle exhaust as part of a highway act or code are listed as possessing no legislation.

Africa (continued)

Paarl (South Africa)⊙ →
Pretoria (South Africa)○ →
Rabat-Salé (Morocco)●□
Salisbury (Rhodesia)● □
Sekondi-Takoradi (Ghana)○
Sfax (Tunisia)●□
Tunis (Tunisia)●□
Umtali (Rhodesia)⊙
Worcester (South Africa)○

Asia and Far East

Bombay (India)○
Cebu (Philippines)● □
Damascus (Syria)●□
Delhi (India)○
George Town, Pinang (Malaysia)⊙
Hakodate (Japan)⊖■
Hiroshima (Japan)⊖■
Hong Kong (China)●□
Kuala Lumpur (Malaysia)⊙ →
Manila (Philippines)●□
Naha (Japan)⊖■
Osaka (Japan)●■⊞→
Shizuoka (Japan)⊖■
Singapore (Malaysia)⊙ →
Tokyo (Japan)■
Yogyakarta (Indonesia)○
Zamboanga (Philippines)⊙

Australasia

Adelaide, S. A. (Australia)⊙
Auburn, N.S.W. (Australia)○
Auckland (New Zealand)●□⊞→
Ballaarat, Vict. (Australia)⊖●□
Bankstown, N.S.W. (Australia)○ →
Bendigo, Vict. (Australia)○⊖
Brighton, Vict. (Australia)⊖⊙⊞→
Brisbane, Qnsld. (Australia)⊙ →
Cairns, Qnsld. (Australia)⊙
Camberwell, Vict. (Australia)⊖⊙⊞→
Canberra, N.S.W. (Australia)○⊞→
Coburg, Vict. (Australia)⊖●⊞→
Footscray, Vict. (Australia)⊖●⊞→
Gold Coast, Qnsld. (Australia)⊙●□■→
Heidelberg, Vict. (Australia)⊖⊙ →
Hobart, Tas. (Australia)⊙ →

Ipswich, Qnsld. (Australia)⊙
Katoomba, N.S.W. (Australia)○ →
Launceston, Tas. (Australia)●□ →
Maitland, N.S.W. (Australia)○
Marion, S.A. (Australia)○
Melbourne, Vict. (Australia)⊖●□■⊞→
Mitcham, S.A. (Australia)⊙
Parramatta, N.S.W. (Australia)○
Penrith, N.S.W. (Australia)○
Perth, W.A. (Australia)○ →
Port Adelaide, S.A. (Australia)●□ →
Richmond, Vict. (Australia)⊖■
Rockdale, N.S.W. (Australia)○ →
Sutherland, N.S.W. (Australia)○ →
Sydney, N.S.W. (Australia)●□ →
Toowoomba, Qnsld. (Australia)● ■
Unley, S.A. (Australia)● □
Waverley, N.S.W. (Australia)○
Wellington (New Zealand)⊙
West Torrens, S.A. (Australia)●□
Whyalla, S.A. (Australia)○
Wollongong, N.S.W. (Australia)○ →

Central and South America

Acapulco (Mexico)○
Campinas (Brazil)● ■
Chiclayo (Peru)●□■
Manizales (Colombia)●□
Mérida (Venezuela)●□■
Ribeirão Prêto (Brazil)● ■
Rio Grande (Brazil)○
San Juan (Puerto Rico)⊙● □
Săn Salvador (Salvador)●□
Săo Paulo (Brazil)●□■

Europe

Amsterdam (Netherlands)●□■
Åarhus (Denmark)⊖● ■
Athens (Greece)⊙□
Biarritz (France) ●□
Birmingham (England)⊖●□
Basle (Switzerland)⊖●□■
Bern (Switzerland)⊖●□■
Bonn (W. Germany)⊙● □
Bordeaux (France)⊖●□■
Brest (France)●
Bydgoszcz (Poland)⊖●
Cologne (W. Germany)●□■

Copenhagen (Denmark)⊖●■
Corfu (Greece)○
Cork (Eire)●○
Dublin (Eire)●○
Essen (W. Germany)●○→
Florence (Italy)●○
Freiburg (W. Germany)⊙
Frankfurt am Main
    (W. Germany)●○■→
Geneva (Switzerland)●○■
Genoa (Italy)⊙
Glasgow (Scotland)⊖●○
Göteborg (Sweden)⊖●○■
Graz (Austria)○ →
Hamburg (W. Germany)⊙●○
Helsinki (Finland)⊖●⊞→
Inverness (Scotland)⊖●○
Izmir (Turkey)●○
Karlsruhe (W. Germany)⊖●○
Kingston upon Hull
    (England)⊖●○
Lausanne (Switzerland)⊖●○■
Leeds (England)⊖●○
Liège (Belgium)○→
Lisbon (Portugal)●○
London (England)⊖●○
Luxembourg (Luxembourg)●○⊞→
Malmö (Sweden)⊖●○
Manchester (England)⊖●○
Monaco●○⊞→
Munich (W. Germany)●○
Nancy (France)⊖○■
Nantes (France)⊖○■
New Lisbon (Portugal)●■
Odense (Denmark)⊖●○■
Oporto (Portugal)●○■
Oslo (Norway)●○■
Paris (France)⊖●○■
Plymouth (England)⊖●○
Saarbrücken (W. Germany)●○
Saint-Nazaire (France)⊖●○■
Sheffield (England)⊖●○
Southampton (England)⊖●○
Stockholm (Sweden)⊖●○■
Stuttgart (W. Germany)●■
Toulon (France)⊖●○■
Turin (Italy)○
Turku (Finland)●○⊞→
Uppsala (Sweden)⊖●○■
Wiesbaden (W. Germany)⊖●○■

North America

Albany, N.Y. (U.S.A.)○ →
Albuquerque, N. Mex. (U.S.A.)●■ →
Anchorage, Alaska (U.S.A.)●■
Atlanta, Ga. (U.S.A.)●○
Austin, Tex. (U.S.A.)●○⊞→
Barrie, Ont. (Canada)●○
Baton Rouge, La. (U.S.A.)●○■
Birmingham, Ala. (U.S.A.)●○
Boston, Mass. (U.S.A.)●■
Brandon, Man. (Canada)⊙
Buffalo, N.Y. (U.S.A.)●○
Burnaby, B.C. (Canada)●○■
Calgary, Alta. (Canada)●○■
Charlottetown, P.E.I. (Canada)⊙
Chattanooga, Tenn. (U.S.A.)●○→
Chicago, Ill. (U.S.A.)●○■
Cleveland, O. (U.S.A.)⊙⊞→
Dallas, Tex. (U.S.A.)●■
Edmonton, Alta. (Canada)●○■
El Paso, Tex. (U.S.A.)● ○
Fairbanks, Alaska (U.S.A.)●○
Fort Worth, Tex. (U.S.A.)●○
Fredericton, N.B. (Canada)⊙
Fresno, Cal. (U.S.A.)●■
Grand Rapids, Mich. (U.S.A.) ●○■
Great Falls, Mont. (U.S.A.)●○■
Halifax, N.S. (Canada)● ○
Hartford, Conn. (U.S.A.)●○→
Helena, Mont. (U.S.A.)●○■
Indianapolis, Ind. (U.S.A.)○→
Jackson, Miss. (U.S.A.)○
Jacksonville, Fla. (U.S.A.)● ○■
Juneau, Alaska (U.S.A.)○
Kansas City, Mo. (U.S.A.)●○→
Little Rock, Ark. (U.S.A.)●○⊞→
Los Angeles, Cal. (U.S.A.)●■
Madison, Wis. (U.S.A.)●■→
Miami, Fla. (U.S.A.)⊙
Milwaukee, Wis. (U.S.A.)●■
Mobile, Ala. (U.S.A.)⊙
Montréal, Qué. (Canada)●○⊞→
Nashville, Tenn. (U.S.A.) ●○⊞→
Oklahoma City, Okla. (U.S.A.)●○
Omaha, Neb. (U.S.A.)●○
Ottawa, Ont. (Canada)●○■
Phoenix, Ariz. (U.S.A.)○
Pierre, S. Dak. (U.S.A.)●○
Pittsburgh, Penn. (U.S.A.)⊙⊞→

North America (continued)

Portland, Ore. (U.S.A.)●□⊞►	Seattle, Wash. (U.S.A.)●□
Québec City, Qué. (Canada)●□■	Sioux City, Ia. (U.S.A.)⊙
Raleigh, N.C. (U.S.A.)●□	Springfield, Ill. (U.S.A.)●□
Regina, Sask. (Canada)⊙►	Sudbury, Ont. (Canada)●□
Rimouski, Qué. (Canada)●□	Tallahassee, Fla. (U.S.A.)○
Saint Augustine, Fla. (U.S.A.)⊙⊞►	Thunder Bay, Ont. (Canada) ●□
St. John's, Nfld. (Canada)○	Toronto, Ont. (Canada)●□■
St. Paul, Minn. (U.S.A.) ●■	Tucson, Ariz. (U.S.A.)⊙►
Salt Lake City, Ut. (U.S.A.)●■	Wichita, Kan. (U.S.A.)○
San Diego, Cal. (U.S.A.)●■	Vancouver, B.C. (Canada)●□
Santa Fe, N. Mex. (U.S.A.)○►	Victoria, B.C. (Canada) ⊙
Savannah, Ga. (U.S.A.)⊙	Winnipeg, Man. (Canada) ●■⊞►

From the list we can see that there is a considerable amount of anti-noise activity on all continents. Much of this interest is recent, as can be seen from the following graph, which gives the total number of pieces of legislation passed each year by the communities listed.

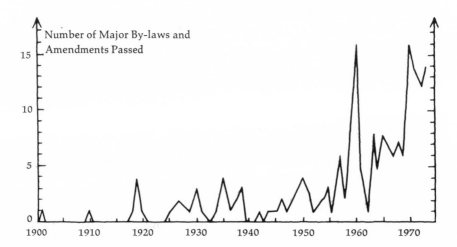

For comparison, here is another graph for ninety communities in Canada, which shows that recent concerns about noise extend also to cities and towns with smaller populations.

At this point it is advisable to point out that cities with no anti-noise legislation should not necessarily be considered backward; perhaps they are just quieter. For instance, while the principal cities of India possess no anti-noise legislation, sound level readings taken in numerous districts of

Bombay at night were all significantly lower than the upper limit advocated by Norway for residential districts at night.*

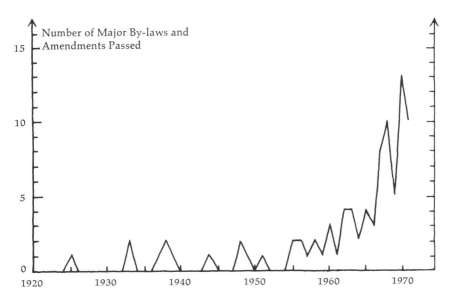

Although community noise legislation around the world is various, certain subjects reappear with predictable regularity:

shouting or creating a disturbance in public;
street and house music;
loudspeakers, radios, etc.;
noisy animals;
unmuffled motor vehicles;
noisy industry in residential areas.

While these are common to many parts of the world, the stress is often different. Thus while in northern countries dogs prove to be a special

Bombay Districts	*Midnight*	*3 a.m.*
Dadar (B.B.)	40 dBA	35 dBA
Ghatkopar	47 dBA	43 dBA
Wadala	35 dBA	30 dBA
Vile Parle (West)	33 dBA	25 dBA
Kalbadevi	50 dBA	45 dBA

The Norwegian level for residential districts is 55 dBA at night during the summer, and 60 dBA during the winter. The levels for Bombay were obtained by S. K. Chatterjee, R. N. Sen and P. N. Saha (see note, p. 198). Most of these levels also fall below those required by Tokyo law, which is 45 dBA for residential districts at night. For further comparison, the night level for Sweden is 40 dBA and that for Richmond (Australia) is 30 dBA.

nuisance, in Latin America it is radios and loudspeakers. In many Latin American cities this is the only item covered by the noise by-law, which is usually aimed at restricting the use of loudspeakers by vendors and outside business establishments. (Venezuela appears to be unique in prohibiting music in busses and taxis if passengers find it objectionable.)

Restrictions against blowing car horns are found in numerous by-laws around the world although the enforcement of the legislation varies enormously. While cities like Tunis restrict the blowing of horns to "un ou deux coups brefs . . . en cas d'absolue nécessité seulement," when we had researchers count the number of horns sounded at main intersections in several of the world's capitals, the Middle East appeared to be the most tolerant of car-horn noise. Here are the hourly averages for a few different cities:

Moscow	17
Stockholm	25
Vancouver	34
Utrecht	37
Toronto	44
Sydney	62
Vienna	64
Amsterdam	87
London	89
Tokyo	129
Boston	145
Rome	153
Athens	228
New York	336
Paris	461
Cairo	1150

These counts were done in 1974–75. Our method was to count all audible horns at an intersection over a period of nine hours on a typical weekday. Of special interest is the fact that previous counts done at the same locations in London and Paris yielded far fewer horns. In fact, the number had about quadrupled in both cities in four years! Could this be part of a larger pattern of increase?

Our civic noise legislation list distinguishes between those communities possessing qualitative and quantitative legislation. It also shows that a considerable number of communities are expecting to adopt or are actively studying some form of quantitative legislation. As this appears to be a trend, at least among technologically developed nations, it calls for a comment. Quantitative legislation shifts the burden of investigation and proof onto the civic administration, which is an encouraging change, for the difficulty with qualitative legislation has always been to prove an

allegation of noise in court. Nevertheless, quantitative legislation is difficult and costly to supervise, for expensive equipment must be purchased
and it must be operated properly by trained staff.

Two approaches to the quantification of sound measurement are currently employed. The first, which is favored in such parts of the world as
Japan, Scandinavia and Australia, is to establish acoustic zones in the
community (residential, commercial, industrial, etc.) and to fix general
sound levels for each zone for particular times of the day and night. A
second method is to set limits for particular offensive noises and to concentrate on keeping them under control. This method is preferred in Canada,
the U.S.A. and parts of Europe and South America. Little standardization
exists from country to country or even from community to community in
the same country. For example, among the seven Canadian cities with
quantitative limits for automobiles in 1972, the permissible level ranged all
the way from 80 decibels at 20 feet (Burnaby, B.C.) to 94 decibels at 15
feet (Toronto, Ontario). In fact, after a careful investigation of quantitative
legislation from many countries, one is forced to conclude that there is
every bit as much capriciousness involved in its formulation as in the other
kind.

The difficulty is that while established levels may be able to cope with
physically destructive sounds, they do not relieve the problem of psychologically distressing nuisances. This is an area in which too little research
has been conducted, and in any case it is hardly likely to lend itself to such
easy quantification. It is for this reason that the most practical anti-noise
legislation for the present era should contain both quantitative and qualitative provisions, though as human nature seems always disposed to an
either-or approach to problem-solving, we will probably witness an ero-
sion of the qualitative in favor of the quantitative, for the latter better suits
the technocratic mind.

How the Study of Noise Legislation Reveals Cultural Differences

Noise by-laws are not created arbitrarily
by individuals; they are argued into existence by societies. Hence they can
be read to reveal different cultural attitudes toward sound phobias. For
instance, along with articles dealing with well-known noise sources, the
city of Genoa (Italy), in its *Regolamento di Polizia Comunale* (1969), identifies some unusual problems. Article 65 states that from 9 p.m. to 7 a.m.,
shutters must be opened and closed as quietly as possible. A keynote for
the European, the shutter is a soundmark for the outsider. In this connection, too, I am reminded that in Étienne Cabet's nineteenth-century utopia,
Voyage en Icarie, the author describes the wonderful invention of noiseless
windows with which each Icarian's house is fitted. Article 67 of the Geno-
ese regulations places restrictions on the noise of furniture moving between 11 p.m. and 7 a.m.—a matter which may cause some puzzlement

until an Italian explains that often heavy work such as moving is carried on at night to avoid the summer heat. Article 70 contains the customary proscription against street music and article 73 explains that Bocci (Bowls) must not be played after midnight. It is surprising to find specific games mentioned as sources of noise, but outdoor bowling is frequently mentioned in the by-laws of European cities. In Luxembourg it is prohibited to "jouer aux quilles" between 11 p.m. and 8 a.m.

A curious piece of legislation, found in Germanic countries but not elsewhere, is directed against the beating of carpets and mattresses. This was encountered in Switzerland (page 190). In Bonn (Germany) "the beating of carpets, mattresses or other objects is only tolerated on weekdays between the hours of 8 a.m. to 12 noon and additionally on Fridays from 3 p.m. to 9 p.m." In Freiburg (Germany) the law is the same but the times are slightly different: weekdays 8 a.m. to 1 p.m. and 3 p.m. to 9 p.m.*

The siesta is a time of reduced energy and many by-laws restrict noisy activities during this period. It is interesting, however, to observe how the siesta expands as one moves south into the stronger sun. In northern Europe it is generally two hours: 1 p.m. to 3 p.m. In Italian cities it often extends from 12 noon to 4 p.m.; but in North Africa it runs to 5 p.m. Typical is this by-law from Tunis: "Between the hours of 10 p.m. and 8 a.m. during the whole year, and between 12:30 p.m. and 5 p.m. from June 1 to September 30 inclusive, it is forbidden to produce or permit to be produced, noises which may disturb the tranquility of the neighbourhood." Interesting are some of the noises identified in this by-law: "This prohibition applies notably to noises produced in the street or on private property by automobile horns, musical instruments, testing motors, exhaust noise, trumpets employed by gasoline salesmen [!], whistles or cries of ice-cream vendors, or any other calls." It was this last sound that Albert Camus recalled especially when he reflected on life in North Africa: ". . . one incident stands out . . . I heard the tin trumpet of an ice-cream vendor in the street, a small, shrill sound cutting across the flow of words."

Certain strong but quite vernacular noises may be considered soundmarks—albeit negative ones. For instance, among the most complained of sounds in the city of Essen (Germany) is the restaurant noise of pounding veal for schnitzels. In Hong Kong a principal source of noise complaints is the sound of "mah jong parties." The slapping together of mah-jong tiles is also characteristic of the Chinese districts of Vancouver and San Francisco, where it is enjoyed by the tourists.

India is becoming concerned about the noise of air-conditioners at night in the fancy tourist hotels now being built, a problem which has already attracted the largest number of public complaints of any noise in

*Freiburg also permits the mowing of lawns only from 8 a.m. to 1 p.m. and 3 p.m. to 8 p.m. Outside the Germanic countries, we have obtained information from only one other city with a by-law against carpet-beating—Adelaide (Australia): By-law No. IX, Paragraph 25b (1934).

the affluent city of Chicago. In places where there are such hotels in India the sound level was found to be 15 to 20 decibels higher at night.

In Mombasa (Kenya), some of the most common noises are "tin-beaters, drum-beaters, blacksmiths and charcoal-stove makers"; while in the port city of Auckland (New Zealand), a major source of complaint is "backyard panelbeating and boat building"—and there is a by-law designed to prevent it from taking place at night. In Rabat (Morocco) one of the chief noises is family reunions, while in Izmir (Turkey) it is unruly behavior at bus terminals (one has to have visited Turkey to appreciate this) and the matter is dealt with in a rather whimsical piece of legislation: "In bus terminals and parking lots proper behaviour is expected. Any case of noise which may disturb the public—for instance, shouting, fighting, etc.—is subject to a penalty. Any claim that the act was a joke will not change the consequence, which is a 50 Turkish Lira fine."

By-laws also reveal the different states of development of societies. While in 1961 the city of Melbourne (Australia) repealed an ancient by-law prohibiting "the ringing of auctioneer's bells," because it was no longer necessary, in the same year Manila (Philippines) found it necessary to introduce legislation against such activity: "No bell or crier or other means of attracting bidders by the use of noise or show, other than a sign or flag shall be employed."

While legislation against loudspeakers and amplified music is common around the world, it is important to note the exemptions worked out by various societies. For instance, Manila, while prohibiting the operation of outside radios and phonographs, retains them for "restaurants, refreshment parlors, beauty parlors and barber shops . . . from 7 a.m. to 12 midnight." Vendors of "newspapers and peddlers of ice cream, fruit, candies, confections and sweetmeats" are also permitted the use of a megaphone or magnavox from 5 a.m. [!] to 11 p.m. Another ordinance in Manila states that the prohibition of amplified sound shall not apply to "any house or place where the sacred passion of Christ is recited or sung during Lent . . . and in cases of important international or national radio events."

Keeping Sunday quiet, which was a special concern in Switzerland, is by no means a concern of all Christian countries. In San Salvador, for instance, the following schedule to control the use of loudspeakers shows that religious holidays are to be regarded as festive occasions. Loudspeakers are permitted at the following times:

Monday to Saturday: 12 noon to 10 p.m.,
Sundays and Holidays: 8 a.m. to 10 p.m.,
December 24 and 31: 8 a.m. to 5 a.m. of the following day.

We have seen from Chicago how church bells are beginning to become a source of complaint—as was predicted earlier in this book. The same sound is also beginning to cause concern in Germany, and in Luxembourg it has been mentioned as a source of growing complaint. In fact a number

of municipalities have already enacted legislation restricting the times
when church bells may be rung. Manila restricts them to no more than
three minutes per hour during the day and prohibits them entirely between
8 p.m. and 6 a.m. In Chiclayo (Peru) they are prohibited between 9 p.m.
and 6 a.m. In Genoa they are to be silenced on complaint and Hartford
(Connecticut) "prohibits the ringing of bells attached to any building or
premises which disturbs the quiet or repose of persons in the vicinity
thereof."

     Such legislation has a parallel in the Muslim city of Damascus (Syria),
where "the reciting of the Qur'an is strictly prohibited on the radio, or by
any person in a public place, particularly restaurants, night clubs and
places of amusement."

Noise in Language   In the days when Christianity was a ruling
force, the church proscribed some sounds of its own. Blasphemy was
rewarded with dire penalties. This notion has often been carried over to
civil noise by-laws, particularly in Anglo-Saxon countries. In Canada such
prohibitions exist in laws in Rimouski and Laval (Québec) and in Brandon
(Manitoba). Similar articles are found in the by-laws of Adelaide (Australia), and in those of such American cities as Buffalo (New York) and
Sioux City (Iowa). Usually the by-law refers merely to "obscene or profane
language," but that of Oklahoma City puts the matter more explicitly: "It
shall be unlawful and an offence for any person . . . to wantonly utter . . .
reproach or profane ridicule on God, Jesus Christ, the Holy Ghost, the
Holy Scripture, or the Christian or any other religion calculated, or where
the natural consequence is, to cause a breach of the peace or an assault."
That such by-laws are still operative, at least in some quarters of the
Anglo-Saxon world, is proven by the fact that in Salisbury (Rhodesia),
1,788 prosecutions were obtained for swearing in the street in 1972.

     In Wellington (New Zealand), an interesting legal anomaly existed
(until it was overcome by new legislation in 1973) which draws our attention to the conflict between physical and acoustic space. For a prosecution
against foul language to be possible, both the offender and the complainant
had to be in a public place. "A person using obscene language in the street
could not be prosecuted by the police if the complainant said he heard the
language while he was standing in private property (e.g. his garden) or
while inside his house. The complainant would have to move out into the
street for a case to be made out." Some day a philologist will write the
international history of four-letter words. What appears to happen is that
with time certain sacred words are debased to become expressions of
public vociferation. It was, for instance, during the decade 1960–70 that the
sacred words "Christ," "God" and "Jesus" found their way in North
America into public conversation as expletives. The release of sacred words
as expletives into the conversational soundscape has one purpose: to

shock. And shock they do, at least until habit mollifies them or they are transformed euphemistically. (In parenthesis one might remark that many sacred words lend themselves well to such harsh usage, for they are often phonetically percussive or jarring. "Jesus" and "Christ" are not exempt from this; the first word scrapes and the second cracks crisply over the tongue.)

But there is much more to the question of blasphemy than first meets the ear, for the fact is that taboo words will always exist in a society. No society has ever had the courage to expose all the dark regions of its psyche to the freedom of daylight and none ever will. Thus, as certain four-letter expressions are released into public garrulity, others take their place as unutterable shockers. The new four-letter words of the English language are "grace," "virtue," "virgin," "tenderness."

Taboo Sounds   I have frequently stated throughout this book that the real value of anti-noise legislation is not the degree of its efficiency—for, at least since the Deluge, it has never been efficient—but rather that it affords us comparative catalogues of sound phobias from different societies and different times. Proscripted sounds thus have enormous symbolic resonance. Primitive peoples guarded their taboo sounds very carefully, and Sir James Frazer devotes a whole chapter of his monumental study *The Golden Bough* to the subject. There we learn that some tribes will not utter the names of certain people, of enemies or dead ancestors, for instance, out of sheer fright. Among other tribes the pronouncement of one's own name may deprive the owner of vital powers. To breathe that most personal sound would be like extending one's neck to the executioner.

Even more interesting, from the point of view of anti-noise enforcement, are customs observed by some tribes which restrict the production of certain noises to certain times out of fear of divine wrath.

> Noises associated with the day are always forbidden at night: for instance, women may not pound grain after dusk. . . . Noisy work seems to bring the village into a dangerous relation with the forest, except on specified occasions. On ordinary days the spirits are sleeping in the farthest depths of the forest, and would not be disturbed, but on the day of rest they come out, and may be near the village. They would be angry to hear chopping sounds in the forest, or pounding in the village.

The Christian habit of observing silence on the Sabbath may have had a similar thought-basis.

Traditionally, taboo sounds, if inappropriately uttered, were always followed by death and destruction. This is true of the Hebrew word *Jaweh,* and it is also true of the Chinese *Huang Chung* (Yellow Bell) which, if sounded by the enemy, would be sufficient to cause the collapse of empire and state. The Arabs, too, had many words for Allah which possessed the same terrible powers (breathe softly as you read them): *Al-kabīd, Al-Muthill, Al-Mumīt,* and ninety-nine others.

Where do we locate the taboo sounds in the contemporary world? One is certainly the civil defense siren, possessed by almost every modern city, held in reserve for that fatal day, then to be sounded once and followed by disaster.

There is a deep-bonded relationship between noise abatement and taboo which cannot be abandoned, for the moment we place a sound effectively on the proscription list we do it the ultimate honor of making it all-powerful. It is for this reason that the petty proscriptions of the community by-law will never succeed, indeed must never succeed. The final power then is—*silence,* just as the power of the gods is in their invisibility. This is the secret of the mystics and the monks; and it will form the final meditation of any proper study of sound.

PART FOUR

Toward Acoustic Design

FOURTEEN

~~~~~~

Listening

Acoustic Ecology and Acoustic Design   The most important revolution in aesthetic education in the twentieth century was that
accomplished by the Bauhaus. Many famous painters taught at the Bauhaus, but the students did not become famous painters, for the purpose of
the school was different. By bringing together the fine arts and the industrial crafts, the Bauhaus *invented* the whole new subject of industrial design.

An equivalent revolution is now called for among the various fields
of sonic studies. This revolution will consist of a unification of those
disciplines concerned with the science of sound and those concerned with
the art of sound. The result will be the development of the interdisciplines
acoustic ecology and acoustic design.

Ecology is the study of the relationship between living organisms and
their environment. Acoustic ecology is therefore the study of sounds in
relationship to life and society. This cannot be accomplished by remaining
in the laboratory. It can only be accomplished by considering on location
the effects of the acoustic environment on the creatures living in it. The
whole of this book up to the present chapter has had acoustic ecology as
its theme, for it is the basic study which must precede acoustic design.

The best way to comprehend what I mean by acoustic design is to
regard the soundscape of the world as a huge musical composition, unfold-
ing around us ceaselessly. We are simultaneously its audience, its performers and its composers. Which sounds do we want to preserve, encourage,
multiply? When we know this, the boring or destructive sounds will
become conspicuous enough and we will know why we must eliminate
them. Only a total appreciation of the acoustic environment can give us

the resources for improving the orchestration of the soundscape. Acoustic
design is not merely a matter for acoustic engineers. It is a task requiring
the energies of many people: professionals, amateurs, young people—
anyone with good ears; for the universal concert is always in progress, and
seats in the auditorium are free.

Acoustic design should never become design control from above. It is
rather a matter of the retrieval of a *significant aural culture,* and that is a task
for everyone. Nevertheless, in provoking this design concern, certain figures have important roles to play. In particular, composers, who have too
long remained aloof from society, must now return to give assistance to
human navigation. Composers are architects of sounds. They have had the
most experience devising effects to bring about specific listener responses,
and the best of them are masters at modulating the flow of these effects
to provide complex and variable experiences which some philosophers
have described as a metaphor for the life-experience itself.

But composers are not yet ready to assume the leadership role in
reorchestrating the world environment. Some are still devoting themselves
with waspish bitterness to a Parnassus of two or three. Others, sensing the
importance of the larger theme of environmental reconstruction, are fumbling ineptly with it, betrayed by inexperience or hedonism. I recall meeting a young Australian composer who told me he had given up writing
music after becoming infatuated with the beauties of cricket song. But
when asked how, when and why crickets sang, he couldn't say; he just
liked taping them and playing them back to large audiences. I told him:
a composer owes it to the cricket to know such things. Craftsmanship is
knowing all about the material one works with. Here is where the composer becomes biologist, physiologist—himself cricket.

The true acoustic designer must thoroughly understand the environment he is tackling; he must have training in acoustics, psychology, sociology, music, and a great deal more besides, as the occasion demands. There
are no schools where such training is possible, but their creation cannot
long be delayed, for as the soundscape slumps into a lo-fi state, the wired
background music promoters are already commandeering acoustic design
as a *bellezza* business.

The Modules for Acoustic Design
A module is a basic unit
to be used as a guide for measuring. In the human environment it is the
human being who forms the basic module. When architects organize
spaces for human habitation, they use the human anatomy as their guide.
The doorframe accommodates the human frame, the stair the human foot,
the ceiling the human stretch. To demonstrate the binding relationship
between architectural space and the human beings for whom it is created,
Le Corbusier made a man with an upstretched arm his modular symbol and
imprinted it on all his buildings.

The basic modules for measuring the acoustic environment are the human ear and the human voice. Throughout this book I have been thumping the theory that the only way we can comprehend extrahuman sounds is in relationship to sensing and producing sounds of our own. To know the world by experience is the first desideratum. Beyond that lie the wonderful exercises of the imagination—the music of the stones, the music of the dead, the Music of the Spheres—but they are only comprehensible by comparison with what we can hear or echo back ourselves.

We know a good deal about the behavior and tolerances of the ear and the voice. When, as today, environmental sound reaches such proportions that human vocal sounds are masked or overwhelmed, we have produced an inhuman environment. When sounds are forced on the ear which may endanger it physically or debilitate it psychologically, we have produced an inhuman environment.

There are few sounds in nature that interfere with our ability to communicate vocally and almost none that in any way pose a threat to the hearing apparatus. It is interesting to consider, for instance, that while the naked voice can be raised to quite a loud level (say about 80 decibels at a distance of a few feet), it cannot be raised in normal human intercourse to a point where it might endanger the ear (over 90 decibels).* In discriminating against low-frequency sounds, the human ear conveniently filters out deep body sounds such as brainwaves and the movement of blood in our veins. Also, the human hearing threshold has been set conveniently just beyond a level which would introduce a continuous recital of air molecules crashing together. The quiet efficiency of all body movements is another stroke of genius. And has anyone speculated on how inconvenient it would be if the ears, instead of being placed on the side of the head, had been placed next to the mouth, where they would have been subjected to close-quarter vocal garrulity and soup-slurping?

God was a first-rate acoustical engineer. We have been more inept in the design of our machines. For noise represents escaped energy. The perfect machine would be a silent machine: all energy used efficiently. The human anatomy, therefore, is the best machine we know and it ought to be our model in terms of engineering perfection.

Contrary to these simple lessons in acoustic ecology, we live in a time when human sound is often suppressed while mechanical jabberware is encouraged. While some of our students were measuring the noise of a downtown construction site in Vancouver, they were entertained by some members of the Hare Krishna sect, an Eastern movement dedicated to the worship of God with song in the streets. In 1971 this group was arrested under the noise abatement by-law, was convicted, appealed the conviction and lost the appeal. This by-law expressly excludes all noise made by construction and demolition equipment—though the students discovered

*From Scarborough, England, comes the news that a British fisherman won what was billed as the World Shouting Competition by raising his voice to 3 decibels at a distance of three meters.

that such noise often ran as high as 90 decibels at precisely the point where
the Hare Krishna singers were arrested. True, singing or hawking in the
streets is frequently annoying; but when it disappears, so does humanism.

Ear Cleaning   The first task of the acoustic designer is to learn how
to listen. *Ear cleaning* is the expression we use here. Many exercises can be
devised to help cleanse the ears, but the most important at first are those
which teach the listener to respect silence. This is especially important in
a busy, nervous society. An exercise we often give our students is to
declare a moratorium on speech for a full day. Stop making sounds for a
while and eavesdrop on those made by others. It is a challenging and even
frightening exercise and not everyone can accomplish it, but those who do
speak of it afterward as a special event in their lives.

On other occasions we prepare for listening experiences with elaborate relaxation or concentration exercises. It may take an hour of preparation in order to be able to listen clairaudiently to the next.

Sometimes it is useful to seek out one sound with particular characteristics. For instance, try to find a sound with a rising starting pitch, or one
that consists of a series of short nonperiodic bursts; try to find one that
makes a dull thud followed by a high twitter; or one that combines a buzz
and a squeak. Such sounds will not be found in every environment, of
course, but the listener will be forced to inspect every sound carefully in
the search. There are numerous other exercises like this in my music
education booklets.*

Sometimes it is useful to document only single sounds in the soundscape in order to get a better impression of their frequency and patterns
of occurrence. Car horns, motorcycles, airplanes can be counted by anyone
with ears, and it is surprising how discriminating one becomes when
isolating one sound from many. Social surveys can also be conducted
simultaneously in which citizens are asked to estimate the number of such
sounds they imagine occur over a given time period. In repeated exercises
of this sort, we have discovered that the imagined traffic is much below
the actual volume—often as much as 90 percent. For instance, when we
asked West Vancouverites to estimate the number of seaplane flights over
their homes in 1969, the average estimate was 8 per day compared with
an actual count of 65. In 1973 the same experiment was repeated in the
same area. This time the average estimate had risen to 16, but the actual
count had also risen to 106. Exercises like this extend ear cleaning to a
wider public. To be reminded of a sound is to think about it; to miss it is
to listen for it next time.

The tape recorder can be a useful adjunct to the ear. Trying to isolate
a sound for high-fidelity recording always reminds the ear of details in the

---

The Composer in the Classroom, Ear Cleaning, The New Soundscape, When Words Sing,
Toronto, 1965, etc.

soundscape that have previously gone unnoticed. Sound events and soundscapes can be recorded for later analysis and if merited can be permanently stored for the future. It goes without saying that only the best tape recorders should be used for this purpose. When we record sounds we provide them with cards giving the following information:

No. _____ Title: _____

Date recorded: _____ Name of recordist: _____

Equipment used: _____ 7½ i.p.s.   mono _____

                _____ 15 i.p.s.   stereo _____

                _____ other   quadraphonic _____

Place recorded: _____ Distance from source: _____

Atmospheric conditions: _____ Intensity: ____dBA

                                                    ____dBB

                                                    ____dBC

Historical observations: _____

Sociological observations: _____

Additional observations: _____

Names, ages, occupations and addresses of local people interviewed:
_____

Sounds threatened with extinction should be noted in particular and should be recorded before they disappear. The vanishing sound object should be treated as an important historical artifact, for a carefully recorded archive of disappearing sounds could one day be of great value. We are currently building such an archive. Our list is very extensive, but a few examples will suffice for illustration.

The ringing of old cash registers.
Clothes being washed on a washboard.
Butter being churned.
Razors being stropped.
Kerosene lamps.
The squeak of leather saddlebags.
Hand coffee grinders.
Rattling milk cans on horse-drawn vehicles.
Heavy doors being clanked shut and bolted.
School hand bells.
Wooden rocking chairs on wooden floors.
The quiet explosion of old cameras.
Hand-operated water pumps.

We train students in soundscape recording by giving them specific sounds to record: a factory whistle, a town clock, a frog, a swallow. It is not easy if the result is to be "clean," without distracting interferences. How often has the novice recordist, sent out to record a "complete" passage of an aircraft, switched off the machine before the sound has dropped totally below the ambiance? Even the life of the more experienced recordist is often hazardous. On one occasion, for example, a small boy had watched our recording team setting up their sound level equipment and tape recorders to measure and record a particular noon whistle. Just as it began, the boy, who had been carelessly left next to a microphone, said: "Is that the whistle you want, mister?"

One of the recordist's biggest problems is to devise ways of recording social settings without interrupting them. The equipment is conspicuous, and in many situations so is the recordist. Peter Huse catches this in a few lines from his poem *Waves.*

> we stagger into a lounge.
>> Bruce in my leather trenchcoat squeaks
>>> and points the way with his goatee as I,
>
>>>> tweedpocket patched with tape,
>>> floppy beret
>> wired with earphones, and gold-heavy
>
>>> Nagra
>>>> digging into my shoulder,
>> cutting two tracks, I
>
>>> angle the mikes in the handset as if
>>>> the machine is off but
>>> the pots are ganged together at 83,
>
>>> it's on RECORD and hidden inside the leather case
>> Scotch 206 crosses the heads onto
>>> the take-up reel and we're getting
>
>> overlapping heart-shapes of late night
>>> fluorescent ferry atmosphere, a blonde siren
>>>> looms toward us.

(Zoom-in jerky, wobbling frame. Engine rumble.
Door swinging. Close-up: her twisted face left
centre looking left. Shuffling, scraping of chairs. A
few slurred voices, hers loudest, grates the most.)

>>> Note bleached hair. Smell her
>> boozy breath. She's drunk and that and hard up.

(Cut to get whole grouping: Tintoretto/home movie
only harsh lights, blue filter. Two men laugh.)

She waves to us, she is singing
"I wanna hol' your han' ..."
and we get it on tape.

A Tourist in the Soundscape   The student of acoustic design
should keep a soundscape diary, constantly noting interesting variations
in sounds from place to place and time to time. The ear is always much
more alert while traveling in unfamiliar environments, as proved by the
richer travelogue literature of numerous writers whose normal content is
acoustically less distinguished. This at least seems to be true of such
authors as Thoreau, Heinrich Heine and Robert Louis Stevenson. Returning from a trip to Rio de Janeiro (1969), an American student was able to
produce a much more vivid account of the Brazilian soundscape than of
the city in which he lived.

Rio de Janeiro	*New York*
Street hawkers	Traffic
Bargaining in the marketplace	Horns of taxis
Live chickens and birds in the markets	Bums on streets
Man going around swatting flies in restaurants	in the Village
Ice being chipped from blocks (no crushed ice)	Busses
Cars and wagons on cobblestones	Subway trains
Street cleaners sweeping by hand	Foreign languages
Strange dial tone, busy signal and ringing	on streets and
of telephones	in restaurants
Predominance of old cars from 40s and 50s	Occasional drunks
Singing and dancing in the streets; music	on streets at
echoing through the whole city from	night
amplifiers (Carnival)	Police sirens
Old hand-operated elevators	
Steam engines in the country	
Total silence in the classroom when	
teacher enters	
No electrical machines in businesses	
and banks	
250,000 people shouting together	
in a stadium	
Cockatoos	
Monkeys	
Cutting of jacaranda	

When one travels, new sounds snap at the consciousness and are
thereby lifted to the status of figures. But the acoustic designer must be
trained to perceive all aspects of *any* soundscape unmistakably, otherwise
how should he be able to adjudicate it properly? How should he be able

to gauge the effect of signals and soundmarks and know the function of keynotes and background sounds?

It is not enough to remain a tourist in the soundscape, but it is a useful stage in the training program. It enables a person to become detached from the functioning environment in order to perceive it as an object of curiosity and aesthetic enjoyment. Like tourism itself, this type of perception is a recent development in the evolution of human civilization. As the American geographer David Lowenthal has written: "Perception of *scenery* is only open to those who have no real part to play in the landscape." Lowenthal illustrates the observation with quotations from Mark Twain and William James.

To Mark Twain's steamboat traveler, the sunset glows eloquently over the rippling silvery water. To the pilot however: "This sun means that we are going to have wind tomorrow . . . that slanting mark on the water refers to a bluff reef which is going to kill somebody's steamboat one of these nights . . . that silver streak in the shadow of the forest is the 'break' from a new snag."

William James, an early tourist in North Carolina, was able to register the defacement of a beautiful forest by the farmers: "But, when *they* looked on the hideous stumps, what they thought of was personal victory. The chips, the girdled trees, and the vile split rails spoke of honest sweat, persistent toil and final reward." To James, however, "The impression on my mind was one of unmitigated squalor. The settler had . . . cut down the more manageable trees, and left their charred stumps. . . . The larger trees he had girdled and killed . . . and had set up a tall zigzag rail fence around the scene of his havoc. . . . The forest had been destroyed; and what had 'improved' it out of existence was hideous, a sort of ulcer, without a single element of artificial grace to make up for the loss of Nature's beauty."

Because of his dependence on visual stimuli, modern man has allowed himself to be led by the tourist industry into believing that tourism consists simply of sightseeing. But the sensitive human being knows that environment is not merely what is seen or possessed. A good tourist inspects the whole environment, critically and aesthetically. He never merely "sightsees"; he hears, smells, tastes and touches. A tourist of the soundscape would demand not *Sehenswürdigkeiten* but *Hörenswürdigkeiten*. With increased leisure all men could become tourists of the soundscape, remembering affectionately the entertainment of soundscapes visited. All it would take is a little travel money and sharp ears.

Soundwalks   A listening walk and a soundwalk are not quite the same thing, or at least it is useful to preserve a shade of distinction between them.

A listening walk is simply a walk with a concentration on listening. This should be at a leisurely pace, and if it is undertaken by a group, a good rule is to spread out the participants so that each is just out of earshot of the footsteps of the person in front. By listening constantly for the footsteps of the person ahead, the ears are kept alert; but at the same time a privacy for reflection is afforded. Sounds heard and missed can be discussed afterward.

The soundwalk is an exploration of the soundscape of a given area using a score as a guide. The score consists of a map, drawing the listener's attention to unusual sounds and ambiances to be heard along the way. A soundwalk might also contain ear training exercises. For instance, the pitches of different cash registers or the duration of different telephone bells could be compared. Eigentones could be sought in different rooms and passages.* Different walking surfaces (wood, gravel, grass, concrete) could be explored. "If I can hear my footsteps as I walk, I know I am in an ecological environment," said a student. When the soundwalker is instructed to listen to the soundscape, he is audience; when he is asked to participate with it, he becomes composer-performer. In one soundwalk a student asked participants to enter a store and to tap the tops of all tinned goods, thus turning the grocery store into a Caribbean steel band. In another, participants were asked to compare the pitches of drainpipes on a city street; in another, to sing tunes around the different harmonics of neon lights.

A series of ingenious soundwalks ought to be of interest to the tourist industry, and it would be of great value also in introducing ear cleaning into schools.

Exercises such as these are the root of the acoustic design program. Yet they require no expensive equipment and they do not camouflage simple acoustic facts with pictures or statistical displays which, being silent, are *not acoustic information.*

When a school of acoustic design worthy of the title finally comes into existence, ear cleaning must be its basic course.

---

*Eigenton is the German word used to refer to the fundamental resonance of a room, produced by the reflection of sound waves between parallel surfaces. It can be located empirically by singing different notes. The room (particularly an empty one) will resonate quite loudly in unison with the voice when the right note is sounded.

FIFTEEN

~~~~~~

The Acoustic Community

Acoustic Space   We have already encountered the conflict between visual and acoustic space. The influence of our visual orientation has not only left its impression on works of art, but even more emphatically in law. Property is measured in physical terms, in square meters or kilometers. Within the territorial limits of property holdings, the owner is permitted to create a desired environment with comparative freedom. When the world was quieter, privacy was effectively secured by walls, fences and vegetation. When visual and acoustic space were more congruous, the latter required no special attention.

Today acoustic space has important environmental and legal implications not fully appreciated. The acoustic space of a sounding object is that volume of space in which the sound can be heard. The maximum acoustic space inhabited by a man will be the area over which his voice can be heard. The acoustic space of a radio or a power saw will be the volume of space in which those sounds can be heard. Modern technology has given each individual the tools to activate more acoustic space. This development would seem to be running a collision course with the population increase and reduction of available physical space per individual.

A property-owner is permitted by law to restrict entry to his private garden or bedroom. What rights does he have to resist the sonic intruder? For instance, without expanding its physical premises, an airport may show a dramatically enlarged noise profile over the years, reaching out to dominate more and more of the acoustic space of the community. Present law does nothing to solve these problems. At the moment a man may own the ground only; he has no claim on the environment a meter above it and his chances of winning a case to protect it are slender.

214

What is needed is a reassertion of the importance, both socially and ultimately legally, of acoustic space as a different but equally important means of measurement. The following historical observations will assist in replanting this notion.

The Acoustic Community

Community can be defined in many ways: as a political, geographical, religious or social entity. But I am about to propose that the ideal community may also be defined advantageously along acoustic lines.

The house can be appreciated as an acoustic phenomenon, designed for the first community, the family. Within it they may produce private sounds of no interest outside its walls. A parish was also acoustic, and it was defined by the range of the church bells. When you could no longer hear the church bells, you had left the parish. Cockneydom is still defined as that area in East London within earshot of Bow Bells. This definition of community also applies to the Orient. In the Middle East it is the area over which the muezzin's voice can be heard as he announces the call to prayer from the minaret.

An interesting example of an acoustic community from the ninth century shows how the Huns constructed their communities in a series of nine concentric fortified circles. "Between these ramparts, hamlets and farmsteads were arranged in such a way that the human voice could carry from one to the other. . . . Between circle and circle all farms and habitations were laid out in such a way that the news of any happening could be conveyed from one to the other by simply blowing a trumpet."

Throughout history the range of the human voice has provided an important module in determining the grouping of human settlements. For instance, it conditioned the "long" farm of early North American settlers, where the houses were placed within shouting distance of one another in case of a surprise attack, and the fields ran back from them in a narrow strip. The acoustic farm may still be observed along the banks of the St. Lawrence River though its *raison d'être* has vanished.

In his model Republic, Plato quite explicitly limits the size of the ideal community to 5,040, the number that can be conveniently addressed by a single orator. That would be about the size of Weimar in the days of Goethe and Schiller. Weimar's six or seven hundred houses were for the most part still within the city walls; but it was the voice of the half-blind night watchman which, as Goethe tells us, could be heard everywhere within the walls, that expressed best the sense of human scale which the poets found so attractive in the small city-state.

A consideration of the acoustic community might also include an investigation of how vital information from outside the community reaches the ears of the inhabitants and affects their daily routine. We had an opportunity to investigate this when we undertook a soundscape study of the French fishing village of Lesconil, on the south coast of Brittany. Lesconil is surrounded on three sides by the sea and is subject to an onshore-offshore wind cycle known as "les vents solaires." Distant sounds are carried to the village in a clockwise sequence, beginning from the north at night, moving to the east and south during the day, and finally to the west in the evening. In the early morning, when the fishermen put out to sea, the Plobannalec church bells and nearby farming noises are heard clearly. By 9 a.m. it is the bells of Loctudy to the northeast; by 11 a.m. the "puffer" buoys off the east coast; then by noon, the motors of the trawlers out to sea at the south. (On a calm day they can be heard up to 12 kilometers away.) By 2 p.m. the western buoys are clearly heard, and by 4 p.m. it is often possible to hear the blowhole at Point de la Torche, 12 kilometers away to the west. If the weather is foggy, the afternoon will bring the sound of the great foghorn at Eckmühl, on the same coast. By evening, the farm sounds return and with them the bells of Treffiagat to the northwest.

This pattern is characteristic mainly of the summer months when the weather is clear and the fishing is good. Variations in it indicate weather changes: for instance, when certain buoys are heard out of sequence, there will be a squall; or when the surf is strong in the west, good weather will follow. Every fisherman and every fisherman's wife knows how to read the nuances of these acoustic signals and the life of the community is regulated by them.

The acoustic community eventually found itself in collision with the spatial community, as evidenced by numerous noise abatement by-laws. This conflict is also recorded in the decline of Christianity when the parish shrank under the bombardment of traffic noise, just as Islam waned when it became necessary to hang loudspeakers on the minarets, and the age of Goethe's humanism passed when the watchman's voice no longer reached all the inhabitants of the city-state of Weimar. (A further sign of the muzzling of Weimar humanism was a nineteenth-century by-law forbidding the making of music unless conducted behind closed windows.)

Modern man continues this retreat indoors to avoid the canceled environments of outdoor life. In the lo-fi soundscape of the contemporary megalopolis, acoustic definitions are harder to perceive. The sound output of the police siren (100+ dBA) may have surpassed the faltering voice of the church bell (80+ dBA), but such an attempt to produce a new order by sheer might is today proving anachronistic, as increased anomie and social disintegration prove. Today, when the slop and spawn of the megalopolis invite a multiplication of sonic jabberware, the task of the acoustic designer in sorting out the mess and placing society again in a humanistic framework is no less difficult than that of the urbanologist and planner, but it is equally necessary. The problem of redefining the acoustic community may involve the establishment of zoning regulations; but to limit it to this, as is common today, is to mistake the trajectories of the soundscape for the property lines of the landscape. Only when the outsweep and interpenetration of sonic profiles is known and accepted as the operative reality will acoustic zoning rise to the level of an intelligent undertaking.

Outdoor Versus Indoor Sounds   Space affects sound not only by modifying its perceived structure through reflection, absorption, refraction and diffraction, but it also affects the characteristics of sound production. The natural acoustics of different geographical areas of the earth may have a substantial effect on the lives of people. For instance, on the Arkansas prairies Thomas Nuttall (1819) remarked that "no echo answers the voice, and its tones die away in boundless and enfeebled undulations." On the other hand, the heavy forests of British Columbia are richly reverberant. "The dense forest around and beyond seemed to echo back the warning tones of the speaker's voice, and as the congregation united their voices in songs of praise, the very trees seemed to lend their cadence in the melody."

Outdoor sounds are different from indoor sounds. Even the same sound is modified as it changes spaces. The human voice is always raised outdoors. If one takes a portable tape recorder from an indoor room outdoors, talking constantly at the same distance from the microphone, the playback volume will register an increase. This results from the higher ambient noise as well as the fact that with decreased reverberation more vocal energy is required to give the sound the same apparent volume. But also psychologically a public place has replaced a private one and there is often an instinctive tendency to be more demonstrative in a public place. We have had occasion to note (see page 64) that people who live out of doors in hot climates tend to speak more loudly than those who live indoors. It is also significant that northern peoples seem more disturbed by noise than southern.

Any sound intentionally uttered within an enclosure is more or less private, more or less connected with the cult—whether this be the cult of the lover's bed, of the family, of religious celebration or of clandestine political plotting. Primitive man was fascinated by the special acoustic properties of the caves he inhabited. The caves of the Trois Frères and Tuc d'Audobert in Ariège contain drawings depicting masked men exorcising animals with primitive musical instruments. One imagines sacred rites being performed in these dark reverberant spaces in preparation for the hunt.

In the Neolithic cave of Hypogeum on Malta (*c.* 2400 B.C.), a room resembling a shrine or oracle chamber possesses remarkable acoustic properties. In one wall there is a largish cavity at eye level, shaped like a big Helmholtz resonator,* with a resonance frequency of about 90 hertz. If a

---

*A Helmholtz resonator is a cavity-type resonator, so constructed that it will vibrate only at a particular frequency. It was developed by the German physicist Hermann Helmholtz in the nineteenth century to analyze the harmonic components of complex sounds.

man speaks there slowly in a deep voice, the low-frequency components of his speech will be considerably amplified and a deep, ringing sound will fill not only the oracle chamber itself, but also the circumjacent rooms with an awe-inspiring sound. (A child or a woman will not be able to produce this effect, the fundamental pitch of their voices being too high to activate the resonator.)

Early sound engineers sought to carry over special acoustic properties like these into the ziggurats of Babylon and the cathedrals and crypts of Christendom. Echo and reverberation accordingly carry a strong religious symbolism. But echo and reverberation do not imply the same type of enclosure, for while reverberation implies an enormous single room, echo (in which reflection is distinguishable as a repetition or partial repetition of the original sound) suggests the bouncing of sound off innumerable distant surfaces. It is thus the condition of the many-chambered palace and of the labyrinth.

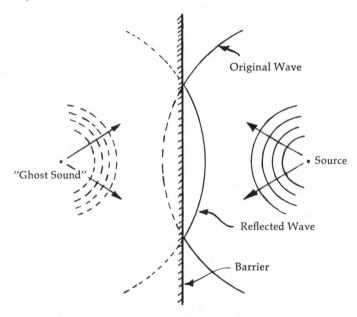

But echo suggests a still deeper mystery. Acousticians will explain that the reflection of a sound off a distant surface is simply a case of the original wave bouncing back, the angles of incidence and reflection being equal. In order to understand this effect one may project a mirror image of the original sound deep behind the surface, at exactly the same distance and angle from the surface as the original. In other words, every reflection implies a doubling of the sound by its own ghost, hidden on the other side of the reflecting surface. This is the world of alter-egos, following and pacing the real world an instant later, mocking its follies. Thus, a far more potent image than Narcissus reflected in the water is that of Narcissus's alter-ego mocking his voice from unseen places behind the rocks. Lucretius, whose philosophy blends science and poetry so skillfully, catches something of this magic quality in his description of the echo:

> One voice is dispersed suddenly into many voices.... I have even seen places to give back six or seven cries when you uttered one: so did hill to hill themselves buffet back the words and repeat the reverberation. Such places the neighbours imagine to be haunted by goat-foot satyrs and nymphs ... they tell how the farmers' men all over the countryside listen, while Pan shaking the pine leaves that cover his half-human head often runs over the open reeds with curved lip, that the panpipes may never slacken in their flood of woodland music.... Therefore the whole place is filled with voices, the place all around hidden away from sight boils and stirs with sound.

Reverberation and echo give the illusion of permanence to sounds and also the impression of acoustic authority. Thus they convert the sequential tones of melody into the simultaneously heard chords of harmony. In open Greek amphitheaters where reverberation was of negligible significance ("never more than a few tenths of a second") harmony was also absent in the musical system. The fact that the theory of harmony was slow to develop in the West was probably due to the way in which Pope Gregory and the medieval theoreticians took over Greek musical theory. Here we have an example of a cultural inheritance inhibiting a natural development, that is, the polyphonic potential of the enclosed forms of Romanesque and Gothic cathedrals. The reverberation of the Gothic church (up to 6–8 seconds) also slowed down speech, turning it into monumental rhetoric. The introduction of loudspeakers into such churches, as has recently happened, does not prove the acoustic deficiency of the churches but rather that listening patience has been abbreviated.

The size and shape of interior space will always control the tempo of activities within it. Again this may be illustrated by reference to music. The modulation speed of Gothic or Renaissance church music is slow; that of the nineteenth and twentieth centuries is much faster because it has been created for smaller rooms or broadcasting studios. This development reached its climax in the highly concentrated, information-packed music of the twelve-tone composers. The contemporary office building, which also consists of small, dry spaces, is similarly suited to the frenzy of modern business, and thus contrasts vividly with the slow tempo of the Mass or any ritual intended for cave or crypt. Now again, the attenuated effects of the newest music seem to suggest a contemporary desire to slow down living pace, just as the music of Stravinsky and Webern had foreshadowed modern business practice.

The Architect as Acoustic Engineer in Antiquity   In a moment I will have some harsh things to say about modern architects' abilities as acoustic designers. But to prepare the case against them it is

necessary to consider them in comparison with their ancient colleagues. Architects of the past knew a great deal about the effects of sound and worked with them positively, while their modern descendants know little about the effects of sound and are thus reduced to contending with them negatively.

The early builders built with ear as well as eye. The exceptional acoustics of the Greek amphitheaters, of which the Asclepius theater at Epidaurus is perhaps the best example, do not prove that acoustics had been totally mastered in ancient Greece, but they do show that a general philosophy of building existed in which acoustic considerations helped determine the form and siting of the structure. In the empty amphitheater at Epidaurus the sound of a pin dropping can be heard distinctly in each of the 14,000 seats—an assertion I have put to the test. That Greek actors were frequently depicted wearing masks with megaphones attached to their mouthpieces does not show that ancient theater acoustics were a failure but merely that Greek theater audiences were probably unruly.

The most beautiful building I have ever experienced is the Shah Abbas Mosque in Isfahan (completed A.D. 1640), sumptuously elegant in gold and azure tile, with its famous sevenfold echo under the main cupola. One hears this echo seven times perfectly when standing directly under the apex of the cupola; standing a foot to either side one hears nothing. Experiencing this remarkable event one cannot help thinking that the echo was no mere byproduct of visual symmetry but was intentionally engineered by designers who knew perfectly well what they were doing and perhaps even used the echo principle in determining the parabolic features of their cupolas.

Something similar apparently exists in the Temple of the Ruler of the Universe in Peking. The actual temple is a circular building, surrounded by a circular wall, inside which are two rectangular buildings, probably indicating the place of the earth within the universe. If a person stands in the center of the site and claps, a series of rapid echoes is heard, caused by reflections from the outer wall. But by moving slightly off-center the echoes will change completely, because only every second reflection will return to the point of origin. In other places near the center, the acoustical conditions are even more complicated, and the echoes will change with even the slightest shift in the placement of the sound-producer. Within this structure it is also possible for persons to converse naturally at great distance when standing just inside the circular wall, for this flat hard-surfaced wall reflects sound around its inner surface with a minimum of transmission loss.

Unfortunately, we have no accounts of how or why acoustic specialties were incorporated into ancient buildings such as these, but since all ancient cultures were strongly auditory, they were quite probably conceived deliberately to express divine mysteries, and at any rate they were certainly not the unpredictable consequences of blueprint accidents. W. C. Sabine, the best architectural acoustician of modern times, studied the "whispering galleries" of some newer buildings: the Dome of St. Paul's Cathedral in London, Statuary Hall in the Capitol at Washington, the vases in the Salle des Cariatides in the Louvre in Paris, St. John Lateran in Rome and the Cathedral of Girgenti. Sabine's conclusion: "It is probable that all existing whispering galleries, it is certain that the six more famous ones, are accidents; it is equally certain that all could have been predetermined without difficulty, and like most accidents would have been improved upon." But these were expressions of a time which was exchanging its ears for its eyes, of a time when the engineering drawing was becoming the prerequisite of architectural thought. This was not so in the Asclepius Theater, in the Shah Abbas Mosque or in the Temple of the Ruler of the Universe. They cannot be "improved upon," for they resulted from the synchronous interaction of *the eye and the ear.*

Among the classical papers on architecture none is more voluminous or respected than the ten books of *De Architectura* by the Roman Vitruvius, which date from about 27 B.C. Book V adequately demonstrates the writer's familiarity with the importance of acoustics, especially in the building of theaters, where, following an extensive exposition of the principles of the Greek science, he discusses the employment of sounding vases in theaters to enhance sound production. Vitruvius writes:

> Hence in accordance with these enquiries, bronze vases are to be made in mathematical ratios corresponding with the size of the theatre. They are to be so made that, when they are touched, they can make a sound from one to another of a fourth, a fifth and so on to the second octave. Then compartments are made among the seats of the theatre, and the vases are to be so placed there that they do not touch the wall, and have an empty space around them and above. They are to be placed upside down. On the side looking towards the stage, they are to have wedges put under them not less than half a foot high. Against these cavities openings are to be left in the faces of the lower steps two feet long and half a foot high. . . .
>
> Thus by this calculation the voice, spreading from the stage as from the centre and striking by its contact the hollows of the several vases, will arouse an increased clearness of sound, and, by the concord, a consonance harmonising with itself.

That these techniques were not special to Vitruvius, we know from the author's own assertion: "Someone will say, perhaps, that many theatres are built every year at Rome without taking any account of these matters. He will be mistaken in this."

These sounding vases were what we now call Helmholtz resonators, and whether or not they originated in Rome, they appear to have been widely used throughout Europe and Asia in the following centuries. They were used in the Shah Abbas Mosque of Isfahan and have been found built into the walls in a number of old Scandinavian, Russian and French churches. In the case of the European churches, the principle appears not to have been completely understood, for the sounding vases do not exist there in sufficient number to produce any noticeable acoustic effect. But a recent discovery of a large quantity of sounding vases (fifty-seven in all) in the small fifteenth-century abbey church at Pleterje, between Ljubljana and Zagreb, shows that the tradition was accurately understood by the Yugoslavian builders, for the double resonance system employed in this case resulted in a high absorption over a broad frequency band from 80 to 250 hertz, an area in which the reverberation time in brick chapels is normally much too long.

From Positive to Negative Acoustic Design   Architecture, like sculpture, is at the frontier between the spaces of sight and sound. Around and inside a building there are certain places that function as both visual and acoustic action points. Such points are the foci of parabolas and ellipses, or the intersecting corners of planes; and it is from here that the voice of the orator and musician will be heard to best advantage. It is here also that the metaphorical voices of sculptured figures will find their true position, not on the metope or tympanum or porch.

Old buildings were thus acoustic as well as visual spectacles. Into the handsome spaces of the well-designed building, orators and musicians were attracted to create their strongest works; there they gained a reinforcement denied them in most natural settings. But when such buildings ceased to be the acoustic epicenters of the community and became merely functional spaces for silent labor, architecture ceased to be the art of positive acoustic design.

In a quiet world, building acoustics flourished as an art of sonic invention. In a noisy world it becomes merely the skill of muting internal shuffles and isolating incursions from the turbulent environment beyond. Thus the great high-rise towers of the world stand on tiptoes, looking out across the fires of the city. *Bellevue—mais mauvais son.*

The Modern Architect as Acoustic Designer   One day I was discussing matters of mutual interest with a group of architecture students. Drawing a picture of a possible future city on the blackboard I asked them what the salient features of this environment appeared to be. There were seven helicopters in the sky of my drawing, yet no student found this particularly salient. I (exasperatedly): "Have you ever *heard* seven helicopters?"

The modern architect is designing for the deaf.

His ears are stuffed with bacon.

Until they can be unplugged with ear cleaning exercises, modern architecture may be expected to continue its same rotten course. The study of sound enters the modern architecture school only as sound reduction, isolation and absorption.

Listen to the sounds a building makes when no one is in it. It breathes with a life of its own. Floors creak, timber snaps, radiators crack, furnaces groan. But although buildings of the past made characteristic sounds, they cannot compete with modern buildings for the strength and persistence of sound emitted. Modern ventilation, lighting, elevators and heating systems create strong internal sounds; and fans and exhaust systems disgorge staggering amounts of noise into the streets and onto the sidewalks around the buildings themselves.

Architects and acoustical engineers have often conspired to make modern buildings noisier. It is a well-known practice today to add Moozak or white noise (its proponents prefer to call it "white sound" or "acoustic perfume") to mask mechanical vibrations, footsteps and human speech. The following thoughts from a recent textbook are typical of the present message being pushed at the graduates and flunctuates of the architectural profession.

Contemporary environmental control can create a complex artificial environment in buildings that will meet all the physical, physiological, and psychological requirements of the occupants. This artificially created synthetic environment is in many respects superior to the natural one. No exterior atmosphere is comparable to an air-conditioned and humidity-controlled room. Lighting fixtures presently available will not only simulate daylight but will create an improved (shadowless) luminous environment indispensable for certain activities.

The author of these comments is Leslie L. Doelle and the book they came from appeared in 1972. Concerning noise suppression Mr. Doelle has this to say:

On the other hand, if the sound is undesirable (noise from a neighbor's television set or traffic noise), unfavorable conditions must be provided for the production, transmission, and reception of the disturbance. Measures must be taken to suppress the intensity of the noise at the source; an attempt must be made to move the noise source as far as possible from the receiver. The effectiveness of the transmission path must be reduced as much as possible, probably by the use of barriers which are adequately sound or vibration-proof, and the receiver must be protected or made tolerant to the disturbance by using noise or background music. All these measures belong to the realm of noise control. . . .

The phenomenon of masking is properly exploited in environmental noise control. If a masking noise is uninterrupted and not too loud, and if it has no information content, it will become an acceptable *background noise* and will suppress other objectionable intruding noises, making them sound psychologically quieter. Ventilating and air-conditioning noises, the noise created by uninterrupted traffic flow of a highway, or the sound of a water fountain are good masking-noise sources.

So much for the memorial drool of Leslie Doelle.

There may indeed be times when masking techniques can be useful in soundscape design but they will never succeed in rescuing the botched architecture of the present. No amount of perfumery can cover up a stinking job.

You are being too severe, the profession insists. In the acoustic design of concert halls and auditoriums, architects and acoustical engineers have brought their work down to a fine science. The fact is that the inventor of room acoustics, Wallace Clement Sabine, still remains, after seventy-five years of this so-called science, its only real luminary. Sabine's Symphony Hall in Boston is still considered probably the finest hall in North America, and it was opened in 1900. Sabine aimed to reproduce the acoustics of the Leipzig Gewandhaus, which has a reverberation time, when empty, of 2.30 seconds. Though the seating capacity of the Boston hall was to be about 70 percent larger than the Leipzig auditorium, he managed to come remarkably close with a reverberation time of 2.31 seconds (empty).

The problem with most modern halls is that they are too large. Here, as in all other aspects of modern life, quantity considerations have forced quality sacrifices. A comparison of some of Europe's best halls (created before the so-called science of acoustics took over) with some pregnant modern structures reveals this clearly enough.

PLACE	DATE BUILT	TOTAL AREA IN SQUARE METERS
Vienna: Grosser Musikvereinsaal	1870	1115
Leipzig: Neues Gewandhaus	1886	1020
Amsterdam: Concertgebouw	1887	1285
New York: Carnegie Hall	1891	1985
Boston: Symphony Hall	1900	1550
Chicago: Orchestra Hall	1905	1855
Tanglewood: Music Shed	1938	3065
Buffalo: Kleinhans Music Hall	1940	2160
London: Royal Festival Hall	1951	2145
Vancouver: Queen Elizabeth Theatre	1959	1975

One of the most spectacular buildings of modern architecture is the Sydney Opera House. The sight of its huge cream-colored butterfly wings, seen from the little, elderly ferries which ply the harbor, is indeed unforgettable, even though the location of the building is convenient rather than inspirational, for the vulgarity of the Sydney skyline behind it and especially the great inelegant bridge at its side, do it no good.

Shortly before it opened in 1973, I was taken on a tour of the Opera House by its sound consultant. I was pleased to note the incorporation of large natural Helmholtz resonators in the walls of the concert hall—functioning more or less exactly as Vitruvius described them two millennia ago —the only hall I know to boast the revival of this technique. In the lobbies, however, I noted the innumerable little speakers which betrayed the inevitable Mooze installation. "The public seems to want it," my guide said feebly.

In the restaurant, the third and smaller but still hugely arched structure of the complex, I was told that the floor was to remain uncarpeted and the kitchen was to be open and situated in the center. I picked up an eight-foot board that was lying at hand and let it fall. The reverberation compared favorably with that of Saint Sofia in Istanbul, probably exceeding eight seconds.

My guide put his finger in his ear and blinked.

If you're ever in Sydney, remember to try out the echo with your soup spoon.

SIXTEEN

~~~~~~

Rhythm and Tempo in the Soundscape

The rhythms of the universe are infinitely various. Some are of such magnitude as to be incomprehensible. Imagine, for instance, that the creation of the world was but one pulse in a great universal symphony of creation and destruction. We do not as yet have an intimation as to when the next pulse may be expected; yet within the incommensurable framework of eternity, these may be but two insignificant cycles contributing the merest fragment of a tone to the universal symphony. Other rhythms are too rapid to be perceived, and can only be designated as "happenings" which, in huge multiplications, give rise to the tiniest recorded events: an instant in the life of a waterfall or a fragment of a radio signal.

Man is an anti-entropic creature; he is a random-to-orderly arranger and tries to perceive patterns in all things. In its broadest sense, rhythm divides the whole into parts. An appreciation of rhythm is therefore indispensable to the designer who wishes to comprehend how the acoustic environment fits together. To do this a scale or module is needed. The possession of such a scale does not imply that all things must be governed by it; only that through it they become more comprehensible. Just as the body of the human being gives us the scale by which architects and designers plan human settlements and by extension can measure uninhabited spaces beyond our reach or control, so the body also gives us modules for comprehending the acoustic rhythms of the environment and universe. What rhythm modules can we discover?

Heart, Breath, Footstep and Nervous System First there is the regular, continuous rhythm of the heart, which may run as low as 50 beats per minute among well-trained athletes, or may go as high as 200 or more during illness or fever, though the normal relaxed beat will be about 60 to 80 per minute. This is a regular dipodic rhythm, pumping in and out, varying only in tempo.

The heartbeat has had a strong influence on the tempo of music. Before the invention of the metronome, the tempi of music were determined from the human pulse, and the difference between a saturnine or frenetic beat in music depended on how far it departed on one side or the other from this modulator. Thus tempi which lie close to the human heartbeat have an obvious appeal for man. In studying the rhythms of Australian aboriginal music, Catherine Ellis discovered that the fundamental drumbeat always hovered near the normal heartbeat. The same has been shown to be the case with Beethoven's "Ode to Joy" from the Ninth Symphony. The composer's original metronome mark of 80 beats per minute is in this range and, while the performance tempi of different conductors vary considerably, the heartbeat range is well respected.

It would be pleasant to conclude that all music or indeed all human activities in this moderate tempo range might be indicative of the well-adjusted society. Unfortunately, this tempo range is also popular among the purveyors of Mooze, where it achieves little more than a bovine character. And by pushing the tempo up slightly, military music has never failed to fire enthusiasm. The heartbeat is nothing more than a rhythm module, roughly dividing humanly perceived rhythms into fast and slow.

Another continuous rhythm is that of breathing, which also varies in tempo with exercise and relaxation. Normal breathing is said to vary between 12 and 20 cycles per minute, that is, 3 to 5 seconds per cycle. But breathing may be slowed down during relaxation or sleep to cycles lasting 6 to 8 seconds. Part of the sense of well-being we feel at the seashore undoubtedly has to do with the fact that the relaxed breathing pattern shows surprising correspondence with the rhythms of the breakers, which, while never regular, often produce an average cycle of 8 seconds.

The correspondence between breath and wave motion was understood by Virgil. In his *Sixth Eclogue* he tells how the Argonauts searched for a lost youth "till the long beach itself called 'Hylas' and again 'Hylas.' " Each cry one breath. Each wave one cry. Perfect synchronization.

The rhythms of all poetry and recited literature bear a relationship to breathing patterns. When the sentence is long and natural, a relaxed breathing style is expected; when irregular or jumpy, an erratic breath pattern is suggested. Compare the jabbing style of twentieth-century verse with the more relaxed lines of that which preceded it. *Something* has happened between Pope and Pound, and that something is very likely the accumulation of syncopations and offbeats in the soundscape. And the perceptible jitteriness in Pound's verse begins after he has moved from rural life in America to the big city of London. Just as human conversational style is abbreviated by the telephone bell, contemporary verse bears the marks of having dodged the acoustic shrapnel of modern life. Car horns punctuate modern verse, not bubbling brooks.

I am surprised that literary critics have not developed the relationship between breathing and writing. Walter Benjamin, at least, has picked up the theme in suggesting that in Proust, a sufferer from asthma, we experience a syntax which suggests fear of suffocating. In his cork-lined study —specially designed to insulate him from city noise—Proust wrote: "The wheezing of my breath is drowning out the sounds of my pen and of a bath which is being drawn on the floor below."

Man also inscribes human rhythms onto the physical world in manual work. Many tasks such as scything, pumping water or pulling ropes cannot be performed unless they follow the breath pattern. Other rhythms— hammering, sawing, casting—take their measure from the arm; still others like knitting or playing a musical instrument are dictated by the motions of the fingers. In instruments like the foot-pedal lathe or loom, hand and foot are united in smooth complementary motions.

Leo Tolstoy describes how Russian peasants could scythe a field together in perfect synchronization without a trace of misdirected energy. Let the description speak for all labor in which body motion, working tool and material are perfectly united.

He heard nothing but the swish of scythes, and saw . . . the crescent-shaped curve of the cut grass, the grass and flower heads slowly and rhythmically falling before the blade of his scythe, and ahead of him the end of the row, where the rest would come. . . .

The longer Levin mowed, the oftener he felt the moments of unconsciousness in which it seemed that the scythe was mowing by itself, a body full of life and consciousness of its own, and as though by magic, without thinking of it, the work turned out regular and precise by itself. These were the most blissful moments.

Another biological tempo which relates significantly to the acoustic environment is that of the resolving power of the sense receptors. In humans this hovers around 16 to 20 cycles per second. It is in this frequency range that a series of discrete images or sounds will fuse together to give an impression of continuous flow. Film employs 24 frames per second in order to avoid flicker. As far as aural perception is concerned, a rapid rhythmic vibration will gradually assume an identifiable pitch at about 20 cycles per second. Thus, as the tempo of human activities increases, the rhythms of foot and hand are mechanized, first into the rough, "grainy" concatenation of the Industrial Revolution's first tools, and finally into the smooth pitch contours of modern electronics. The resolving power of the senses makes it possible to turn some of the nervous agitation of the soundscape into drones which, being less turbulent to the ears, tend to have a pacifying quality.

Within the framework of our experience, the audible metrical divisions of heart, breath and foot, as well as the conservational actions of the nervous system, must be our guide against which we arrange all the other fortuitous rhythms of the environment around us.

Rhythms in the Natural Soundscape The environment contains many sets of rhythms: those dividing day from night, sun from moon, summer from winter. Though these may not provide audible pulsations, they do have powerful implications for the changing soundscape.

To all things there is a season. There is a time for light and a time for darkness, a time for activity and a time for rest, a time for sound and a time for not-sound. It is here that the natural soundscape provides a clue, for if we could register all the periods of rest and activity among natural sounds, we would observe an infinitely complex series of oscillations as each activity rises and falls from exertion to slumber, from life to death. I have put together a rudimentary chart of some of the more prominent natural features of the British Columbia soundscape in order to suggest the shape of the annual cycle. Like that of every other part of the world, it creates a vivid vernacular composition following a general cyclic rule, a composition in which each instrumentalist knows when to perform and when to listen quietly to the themes of others.

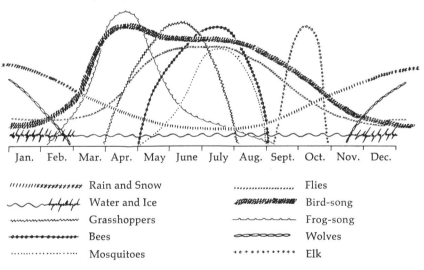

| Jan. | Feb. | Mar. | Apr. | May | June | July | Aug. | Sept. | Oct. | Nov. | Dec. |

Rain and Snow		Flies
Water and Ice		Bird-song
Grasshoppers		Frog-song
Bees		Wolves
Mosquitoes		Elk

The cycles of the natural soundscape of the west coast of British Columbia showing the relative volume of sounds.

Man plays his role in this cycle also, or at least he did when he respected the agrarian calendar. Planting and harvesting contributed rich patterns of seasonal sounds to the country soundscape. And in man's activities too, there were periods of sound and silence—for we were then better listeners, and the forest and field provided acoustic clues vital to survival.

This healthy give and take between sounds in the natural soundscape is disappearing from the modern urban world. When the factories of the Industrial Revolution swivel-moored workers to the same bench for a lifetime, seasonal variations vanished. The factory also eliminated the difference between night and day, a precedent which was extended to the city itself when modern lighting electrocuted the candle and night watchman. If we were to make a continuous recording on a downtown street of a modern city, it would show little variation from day to day, season to season. The continuous sludge of traffic noise would obscure whatever more subtle variations might exist.

Let me instead analyze a recording we made at a rural site near Vancouver. The recording was made by a small pond over a twenty-four-hour period on summer solstice. The accompanying diagram shows the circadian cycle clearly, and sound level readings, taken while the recording was in progress, show the general ambient sound level together with levels of some of the predominant sounds recorded. The loudest of the continuous sounds was that of aircraft, and charts we made during the recording show that this sound was present over the rural setting for an average of 32 minutes per hour during the day and evening.

Aside from the aircraft, there were three main groups of performers: birds, bullfrogs and chirping frogs. The most salient—and for the recordists, the most beautiful—moments of the recording came during the complementary crossovers at dawn and dusk between the chirping frogs and the birds. At the very moment when the first bird was heard (3:40 a.m.) the chirping frogs became silent, not to return until dusk, when the last bird was trailing off. We can have no idea of the significance of this for the performers; we can only note that as the voices of both sets of performers occupy a similar high-pitched range, they would, if sounded together, tend to mask one another, thus reducing the clarity obtained by vocalizing in rotation. The bullfrogs, on the other hand, with deep voices, offer no competition for either performer. They continued to croak intermittently through both day and night.

A similar restraint is noticeable in the gathering of the dawn chorus. Following the first bird, the chorus gradually rose in complexity and intensity, peaking after about half an hour, then settled back to sustain itself on a less frenzied level throughout the day. Of special interest to us was the manner in which each species woke up as a group, sang vibrantly for a few moments, then seemed to fall back as another species came forward. The effect is analogous to the different sections of an orchestra entering separately before combining in the total orchestration. I have found no ornithologist to provide an explanation for this effect, though it is very clear on our recording.

The long-term shape of the other groups of performers is similar, swelling in volume, then decreasing again to inactivity; but in each case this oscillation takes place over a different period of time: about 5 hours for the chirping frog group, about 18 hours for the birds and 24 hours for the bullfrogs.

The two frog groups achieve their peak activity by different means. In the bullfrog group the number of frogs remains constant and each

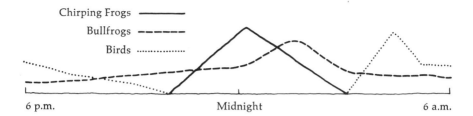

Chirping Frogs ─────
Bullfrogs ─ ─ ─ ─
Birds ·············

6 p.m.   Midnight   6 a.m.

individual frog increases or decreases its vocal activity. In the chirping frog group each individual frog shows little variation in vocal activity but the climax is reached after midnight when the number of active frogs is greatest.

The short-term time pattern of the two frog groups shows a further difference. The chirping frogs sing together in choral interludes separated by silent periods. One or two frogs will start, and almost immediately all will join in. The synchronization with which, after a moment or two of busy activity, they will all stop together is even more remarkable. The bullfrogs, on the other hand, do not operate as a group at all; each frog croaks quite independently, though at the period of their peak activity, their sounds may overlap in ragged rhythmic clusters.

The sounds of frogs and birds are not the only ones on the recording, and though the others do not lend themselves to such rapid analysis, they too have their own unique rhythmic patterns. To make this clearer we mixed an abbreviated tape of the summer solstice recording, selecting two minutes out of each hour, and in this the circadian rhythms can be heard very clearly. We have used the same technique on numerous other occasions in different locations, and have come to regard it as one of the greater learning experiences of recent years.

Rhythms in Village Life   Circadian and seasonal rhythms are observable in human settlements also, but they are strongest in small towns and villages where life is more apt to be regulated by common activities. It was in order to study the dynamics of the village soundscape that we undertook an investigation of five European villages in 1975. Here we were able to document how village life revolves around important community signals such as church bells or factory whistles. We discovered, for instance, that not only do they punctuate village life but their advent precipitates chains of other sounds in quite orderly recitals. For instance, an early morning factory whistle will be preceded by commotion in the streets and followed by commotion in the factory and stillness in the streets. By walking through the various parts of the village several hours each day and making lists of all the sounds heard, we were later able to show the way in which many sounds followed definite rhythmic patterns: for example, the women's voices dominate the streets at certain times of

the day and men's or children's at other times. It was also possible to show
how the rate of traffic growth in a village stimulates a corresponding
growth in the other sounds, but, surprisingly, a *reduction* in the variety of
sounds heard—an important finding which supports earlier statements in
this book. It is difficult to illustrate these matters without elaborate statistical charts, so it is best to refer the interested reader to the study itself,
passing on to make our point that village soundscapes have highly pronounced rhythmic patterns by means of a simpler illustration.

Cembra is an Italian mountain village north of Trento and just below
the Tyrol. Tucked far up a valley, connected to the outside world by a
single, winding, mountain road, Cembra was the only village we studied
in which the number of human sounds in the streets outnumbered those
of motorized traffic. Until well into the twentieth century, Cembra was
virtually self-supporting, producing its own food, goods and services. As
a result it developed a highly active and self-sustaining social life. Entertainments, church feasts and other activities were plentiful and were
strongly acoustic in character.

Winter was the quiet season but was by no means devoid of its
ceremonies. On St. Lucia's and St. Nicholas's Day (December 5) boys
would go around the village ringing hand bells and banging things with
chains. Every so often they would stop to sing the verse of a song about
the saints. On Christmas Eve they returned to sing carols in the streets. At
11:55 p.m. on New Year's Eve, a special bell was rung, echoed the next day
with the firing of cannons, or *mortaretti,* as they are called. These were
small, 15-centimeter-diameter weapons and they seem to have been fired
off on almost all possible occasions and certainly at all religious festivals.
I have mentioned before the intimate relationship between the church bell
and the cannon and how throughout European history the same metal was
poured into one shape, then into the other. With the *mortaretti,* the relationship is made explicit once again.

The relative quiet of winter was shattered on the evening of March
1 with a custom called *Il Tratto Marzo,* when crowds of youths would
climb to different peaks in the hills behind the village. There they would
divide into groups, light fires and, using cardboard megaphones, call names
of those likely to be married in the coming year. If the marriage was a real
possibility, cannons were fired. If the match was a joke, horns would be
blown instead.

During Holy Week the church bells were silenced and ratchets were
used to announce services. Dating back to a pagan period, some of these
ratchets were enormous and were pushed through the streets like wheel-
barrows. Easter used to feature a passion play in the streets. The peasants
still tell the visitor how it was stopped in 1821 when the man portraying
Christ stepped on some spiny chestnut husks, which a few prankish boys
had placed in his path, and cursed so loudly that the bishop put an end
to the procession altogether. When we were there, Easter Day featured
only a procession by the volunteer fire brigade, all decked out in cere-

monial uniforms with epaulets and swords. Afterward they returned to the fire hall and sounded the sirens.

On Easter Day, the bells of Cembra were rung once again, *El Campanò* style, and the *mortaretti* were fired over the valley. The *El Campanò* ring is reserved for special days and consists of a single bell, followed by the ringing of all the bells of Cembra's three churches simultaneously, an effect which is supposed to be beautiful at a distance, say across the valley.

In summer everyone would change their winter hobnailed boots for wooden clogs, and the sound of footsteps—always distinct in Cembra's cobbled streets—changed from metal crunching to hollow wooden clops. Every day the village would hear the goatherd's horn when he took the animals to pasture each morning and returned in the evening. Summer evenings were also a time for group singing. Men, women and children would gather outdoors after supper and sing antiphonally in groups. One special singing event was the *Canta dei Mesi* (Song of the Months) when people would dress in costume and sing the different verses of the song.

The principal summer feasts (St. Peter's and St. Paul's, San Rocco's and the Assumption of the Virgin) were followed by All Saints' Day. Then the cattle were brought back from summer pasture, bells ringing, and the village began to move indoors again. It was then that firewood was hauled from hills above the village in carts which dragged over the cobblestones like skids—a special construction of the area, to help brake the vehicles on the steep hillsides, and a sound which schoolchildren told us was one of their favorites.

I have used the past tense in this brief sketch because so many of the sounds are now no longer to be heard in Cembra and they can only be surmised by discussion with the inhabitants. Today, juke boxes have moved into Cembra, so have big new mountain busses and television sets. Yet Cembra is still a place where one can hear a sound like that of the sacristan, a portly young man, huffing and puffing after ringing the evening bell as he climbs onto his trembling bicycle and spins off into the darkness.

The Rhythms of Radio Broadcasting The modern city does not display such deliberate acoustic rhythms as the village or the natural soundscape. Better stated, the great profusion of rhythms cancel one another out. The principal feature of the city soundscape is random motion and it may be heard to best advantage from a distance or late at night. It is the continuous low-frequency roar one hears from an adjacent hill or through an open window in the small hours of the morning. This is pedesis, Brownian motion, Gaussian noise. It is composed by a million Mr. Browns and Ms. Smiths running around in their private circles or slipping through some more haphazard routines, rarely synchronizing their activities, rarely considering one another.

We do not have enough ceremonies in our lives, Margaret Mead somewhere says, which might be paraphrased to mean that modern social life lacks rhythmic definition. Through a superfluity of activities even special events become monotonous and uniform. Let me illustrate this problem with reference to a single activity which tries to be special all of the time: radio broadcasting. (Television would do as well but since our concern is with audio culture, let's stay with radio.)

While sociologists have done a certain amount of content analysis, a study of broadcast rhythms appears never to have been undertaken. Such research would have special advantages for soundscape studies. First, broadcasting is separated into independent information channels so that the confusion of simultaneity, so often present in the soundscape at large, is absent. Second, broadcasting is a deliberate attempt to regulate the flow of information according to human responses and information-processing capabilities. Third, broadcasting is always changing and these changes can thus be studied, exactly as a critic or historian studies the styles and tendencies of contrasting schools of literature or music. We have already mentioned how broadcasting was at first occasional, separated by extensive periods of silence or low-keyed activity. Here I am concerned only with the shapes of contemporary radio in North America, which today provide the models which much of the rest of the world is copying.

Each radio station has its own style of punctuation and its own methods of gathering the material of its programs into larger units, just as the phrases of language are shaped into sentences and paragraphs. Different events are repeated periodically in daily or weekly schedules, and within each day certain items may be repeated several times at fixed intervals. Station calls are such methods of punctuation. In North America these are repeated frequently, often with musical accompaniment; in other parts of the world they still tend to be more stately, synchronizing with the hours of the clock. In private broadcasting systems, commercials may also be repeated without variation at fixed points throughout the day. These patterns may be called *isorhythms*. Like community sound signals, they anchor time and help listeners to gain temporal orientation. Isorhythmic occurrences may have varied or unvaried content; thus, while certain recorded messages may be repeated exactly, other items, such as the news, may occur at fixed times but use fluctuating material. In news broadcasts we witness a continual evolution of content as units are gradually altered or dropped to make way for new events. In the case of a key news story we may witness a theme with variations spread over several days or weeks.

There are other items which recur unpredictably, but insistently enough to be called *leitmotivs*. For instance, our studies of Vancouver radio stations show that during a one-hour period one disc jockey repeated certain items with great insistence. The name of the station was mentioned 28 times during the hour; the disc jockey mentioned his own name 16 times and the city of Vancouver 13 times during the same period. Following behind these primary citations was a long string of words receiving three or more repetitions. In decreasing order these were: "Travel, first, new, modern, contemporary, popularity, money, perfection, best, reward, prize, convenience, speed, reliability, simplicity, power, entertainment, great, love." During the same hour, 12 records were played and there were 16 commercials. This pattern varies little from day to day, month to month, with the number of commercials—always the maximum permitted by broadcast regulations—eventually emerging as the most ubiquitously recurrent element.

The tempo of broadcasting also deserves careful study. The question to bear in mind here is whether such tempi attempt to reproduce the rhythms of social life, or whether they attempt to alter them by speeding them up or slowing them down. In an extensive word-per-minute count of newscasters on four Vancouver radio stations we were able to determine that reading speeds varied considerably.

CKLG	177.5 words per minute
CBC	184.0 words per minute
CJOR	190.0 words per minute
CHQM	212.3 words per minute

Curiously enough, CKLG is a teenagers' station, concentrating on pop music, while CHQM styles itself as a "relaxed" station for the middle-aged. All of these word-per-minute counts were in excess of those we counted for human conversation, except when it became highly excited.

A 16-hour continuous survey of the same four stations showed something even more revealing: new program items appeared on the average every 1 to 2 minutes—that is to say, this was the amount of time given to a single subject before it was abandoned in favor of a new topic. This is even more brief than the normal pop recording, which long ago fixed the average human interest span for music at no more than three minutes.

The graphs from which I extracted the foregoing information show that the recurrence of isorhythmic and leitmotiv material forms tight little loops throughout the day. It is rarely more than an hour between news broadcasts, rarely more than fifteen minutes between weather reports and rarely more than five minutes between commercials on most of the stations. The all-at-once quality of modern broadcasting is further accented by the uniform dynamic level of programs, a technique known as *compression*. On some stations all program material is compressed to the highest permissible level.

The continuous sound wall of radio broadcasting stands in sharp contrast to the other rhythms mentioned in this chapter. It contradicts them and in many ways has contributed to a lessening of our appreciation for them. This need not be so. Radio, like art, is deliberately created. But art is a skillful selection of experiences, fashioned to give us an intimation of higher, or at least alternative, modes of existence. Radio, too, could be employed to show us alternative modes of living. If modern life is too fast-paced, radio might find a new vocation if, instead of increasing its tempo as it appears to be doing, it assisted man in slowing down by reinforcing once again the natural rhythms of life.

This is the idea behind Bruce Davis's Environmental Radio. Davis's plan is to have microphones installed in wilderness areas from which would be transmitted nothing more or less than the natural soundscape, bringing city-dwellers direct relays from the wilds. The key to the plan is that nothing would be added or edited. The station would simply transmit continuously the sounds at the microphone location.

For years man has been pumping his affairs out across the wildnerness environment. For once the natural soundscape would be allowed, in its wisdom, to speak back to us.

SEVENTEEN

~~~~~~

The Acoustic Designer

When the rhythms of the soundscape become confused or erratic, society sinks to a slovenly and imperiled condition. That was the thesis announced in the introduction to this book. But the other thesis under which I have written is that the soundscape is no accidental byproduct of society; rather it is a deliberate construction by its creators, a composition which may be as much distinguished for its beauty as for its ugliness. When a society fumbles with sound, when it does not comprehend the principles of decorum and balance in soundmaking, when it does not understand that there is a time to produce and a time to shut up, the soundscape slips from hi-fi to lo-fi condition and ultimately consumes itself in cacophony.

It is important to realize that the lo-fi state is not a natural corollary to higher density living or population increase. A visit to the bazaars and traditional towns of the Middle East will impress one by the quiet, almost furtive manner in which large numbers of people manage to go about their business without disturbing one another. The sound sewer is much more likely to result when a society trades its ears for its eyes, and it is certain to result when this is accompanied by an impassioned devotion to machines.

If the acoustic designer favors the ear, it is only as an antidote to the visual stress of modern times and in anticipation of the ultimate reintegration of all the senses.

The Principles of Acoustic Design   The acoustic designer may incline society to listen again to models of beautifully modulated and balanced soundscapes such as we have in great musical compositions. From

these, clues may be obtained as to how the soundscape may be altered, sped up, slowed down, thinned or thickened, weighted in favor of or against specific effects. The ultimate endeavor is to learn how sounds may be rearranged so that all possible types may be heard to advantage—an art called orchestration. The outright prohibition of sound being impossible, and all exercises in noise abatement being consequently futile, these negative activities must now be turned to positive advantage following the indications of the new art and science of acoustic design.

Acoustic design does not, therefore, consist of a set of paradigms or formulae to be imposed on lawless or recalcitrant soundscapes, but is rather a set of principles to be employed in adjudicating and improving them. In addition to the lessons taught by music, these principles consist of:

1. *a respect for the ear and voice*—when the ear suffers a threshold shift or the voice cannot be heard, the environment is harmful;
2. *an awareness of sound symbolism*—which is always more than functional signaling;
3. *a knowledge of the rhythms and tempi of the natural soundscape;*
4. *an understanding of the balancing mechanisms by which an eccentric soundscape may be turned back on itself.*

This last point is most easily understood by turning to Chinese philosophy and art. It is the natural alteration of events that forms the secret of the *yin* and *yang* exchange, the perfect oscillation in which each part implies the existence of the other. Lao-tzu says: "Gravity is the root of lightness; stillness, the ruler of movement." A Chinese painter puts it this way:

Where things grow and expand that is *k'ai;* where things are gathered up, that is *ho.* When you expand (*k'ai*) you should think of gathering up (*ho*) and then there will be structure; when you gather up (*ho*) you should think of expanding (*k'ai*) and then you will have inexpressible effortlessness and an air of inexhaustible spirit.

In ancient Chinese society balance and regulation were highly prized in all things. Excesses of any kind were to be avoided. In the music of this period, *p'ing,* or level unmodulating pitch, with its attributes of smoothness and repose, was contrasted with *tsê,* sudden or contrary movement, with its attributes of assertiveness, activity and aggression. Analyses of pieces of music of the period show that balance was strictly maintained between the two states, so that a composition contained an identical number of *p'ing* and *tsê* features. By contrast, Western music is unbalanced; it is always inclined to be more static or more active. And the soundscape of the West also runs to extremes. There are numerous states of imbalance in need of attention. In each case the term in the left-hand column appears to be dominating that in the right-hand column.

Sound/Not-sound
Technological sounds/Human sounds
Artificial sounds/Natural sounds
Continuous sounds/Discrete sounds
Low-frequency sounds/Mid- or high-frequency sounds

Thought must now be given to how the weighting of these terms may be readjusted to create new harmony and equipoise. These are enormous issues beyond the abilities of any individual to appreciably alter. But the designer does not redesign a whole society: he merely shows society what it is missing by not redesigning itself. And if he does this with passion and talent, his recommendations will eventually be heard and understood. Society is always incapable of imagining improvements without the voice from beyond. Ask Mr. Smith or Mr. Jones what kind of house they want to live in and they'll have you design a hovel every time. It devolves on the designer to point out alternatives.

This is the function of art: to open out new modes of perception and to portray alternative life styles. Art is always outside society and the artist must never expect to win popularity easily. The mind of the designer will move in huge unrealistic excursions too; but he may also engage in some very practical preservation and repair work.

The Preservation of Soundmarks   One practical task of the acoustic designer would be to draw attention to soundmarks of distinction and, if there is good reason to do so, to fight for their preservation. The unique soundmark deserves to make history as surely as a Beethoven symphony. Its memory cannot be erased by months or years. Some soundmarks are monolithic, inscribing their signatures over the whole community. Such are famous church or clock bells, horns or whistles. What would Salzburg be without its Salvatore Mundi, Stockholm without its Stadhuset carillon, London without Big Ben?

In Vancouver, for instance, we have a cannon, built in 1816, which since 1894 has been fired over the harbor each evening, originally to tell fishermen the time, now preserved as a sound souvenir. We also had a diaphone foghorn at the Point Atkinson lighthouse which dated from 1912 and was recently replaced by the Ministry of Transport in its automation scrabble. A more recent arrival (1972) is a set of air horns on top of one of the city's higher buildings which barks out the opening phrase of the National Anthem each day at noon (108 dBA a block away).

Whatever one may think of such soundmarks, they reflect a community character. Every community will have its own soundmarks, even though they may not always be beautiful. For instance: "During the very early period of gold mining [in Ballaarat, Australia] the very many quartz batteries operating caused a constant and sustained noise throughout the whole City area. This was accepted as part of the process of gold extraction."

Some unusual sounds receive legal protection. Thus in the hot city of Damascus the sound of ice-making equipment is specifically exempted from the list of proscripted sounds mentioned in the noise abatement by-law, because such equipment performs a desirable community service and therefore presumably has an attractive symbolism.

It is the less ostentatious soundmarks that need the special vigilance of the acoustic designer, for despite their originality or antique charm, they are more likely to be unceremoniously excised from the soundscape. Often it will take the visitor to point out the value or originality of a soundmark to a community; for local inhabitants it may be an inconspicuous keynote. Let me mention a few originals from my own memory.

- the scraping of the heavy metal chairs on the tile floors of Parisian coffee-houses;
- the brilliant slam of the doors of the old carriages of the Paris Métro, followed by a sharp click, as the latch falls to the locked position (the effect can now, 1976, be heard only on the Marie d'Issy–Port de la Chapelle line);
- the sound of the leather straps on the trams in Melbourne, Australia —when they are tugged they twist around the long horizontal support poles and make rich squeaking noises;
- the virtuoso drumming of the Austrian bureaucrats with their long-handled rubber stamps: *ta-te-te-daa-ta-te-daa;*
- the high-pitched brilliant bells of the horse-drawn taxis in Konya, the last to be heard in any major town in Turkey;
- or in London, the memorable voice on the recording at certain suburban tube stations that says (or used to say), "Stand clear of the doors!"

The world is full of uncounterfeiting and uncounterfeitable sound souvenirs such as these, indelible memories for the aurally sensitive tourist, and always in need of protection against replacement by duds from multinational factories.

Repairs to the Soundscape   Once acoustic design is established as a useful profession, and young designers move out into positions in government and industry, they will be able to effect numerous practical repairs to the soundscape. They might start by correcting some of the bungled design work of their tone-deaf predecessors.

Consider, for instance, a sonic signal for traffic crosswalks. There are several of these already in existence in different parts of the world. I have heard at least three: in Auckland (New Zealand), Växjö (Sweden) and London (England). In Auckland, the traffic lights have three modes: one for vehicular traffic in each direction and one for pedestrians, allowing them the freedom of the intersection. The cue for the pedestrians is a special light plus sonic signal, a hideous buzz, yet neither loud enough nor of sufficiently high frequency to clear the traffic noise. The signals in both London and Växjö consist of rhythmic effects. In London a series of rapid pips, a semitone apart, sound for about four seconds to signal the pedestrian to cross. In the signal I listened to, the two sounds always got out of phase, the one lagging slightly behind the other, in a manner that defied one to decide whether it was a stroke of British design genius or technical ineptitude. In Växjö a single ticking noise has two modes: a fast to go and a slow to stop. The first sounds like a ratchet and the second is set at about the heartbeat tempo of an Olympic sprinter, which is, I suppose, the image intended to be evoked.

The point I am trying to make is that while the acoustic crosswalk is probably a useful idea (one can appreciate its value for the blind), none of the signals heard so far meet all the design criteria, which are both social *and aesthetic.*

North America was introduced to a whole line of obnoxious buzzings when, about 1970, designers decided to equip cars with safety belts and installed a noise to remind the passenger to use them. Here again, the identical question: why must an acoustic attention-getter always be abominable?

Recall the aberrational quality of schizophonic devices such as public address systems. The sudden impact and startle effect of a voice barking an announcement down your neck, exaggerated by the instantaneous switch-on time of electroacoustical devices, is common enough today, but it needs to be redesigned. The technique of attracting attention without frightening the public out of its wits calls for subtle creative action. Sometimes a bell or buzzer is used for this purpose but the envelope of the bell is wrong. The P.A. sound cue should not have a sudden but a sloped attack. Not ⟍ but ⟋ .

The New Zealand Railway employs a tape-loop prelude to P.A. announcements consisting of an eight-second tune on a glockenspiel. But eight seconds is probably too long for repetitive use. In Holland, a short three-tone tune is played on an electric glockenspiel, but this is perhaps too short and boring. The student of acoustic design could profitably be given the assignment of designing a few short preludes to improve such situations.

These examples are only a few of those requiring attention, from a list long enough to keep an army of intelligent people amply occupied.

The Bad Pun of Bell's Telephone Bell   All public announcements abbreviate thought. An intelligent society would keep them to a minimum. The telephone also abbreviates thought. At any moment—perhaps even before I complete my next sentence—a voice from California or London or Vienna may jump onto the table announced by what Lawrence Durrell in *Justine* called "a small needle-like ring."

Who invented the telephone bell? Certainly not a musician. Perhaps it is just a bad pun on the name of its inventor? It may be that such an audacious device should have an obnoxious sound, but the matter should be accorded more consideration. If we must be distracted ten or twenty times each day, why not by pleasant sounds? Why could not everyone choose his or her own telephone signal? In a day when cassettes and tape loops are cheap to manufacture this is entirely feasible.

It is true that the rings of telephones vary considerably from nation to nation. North American phones have single rings, caused by a mechanical clapper vibrating between two bells of identical or near-identical frequency. On Vancouver telephones this ring is repeated every 6 seconds with approximately 1.8 seconds of ring followed by 4.2 seconds of silence.

Although the intensity of the ring on North American telephones can be regulated to some extent by a dial on the base of the instrument, the intensity of the voice speaking and the ringback sound (i.e., the ringing heard over the receiver when a call is placed) are not strictly regulated. In one of our research projects we have registered a busy signal at over 120 dBA and conversation at over 100 dBA at the point where the ear would normally make contact with the receiver. This is loud enough to be an aural health hazard.

A modicum of design consideration was given to the British telephone, for its two loud rings, followed by a pause, was intentionally constructed to total five units—two beats of rings followed by three beats of silence—because it was thought that this asymmetrical meter would be more attention-getting than a meter of three, four or six. New Zealand telephones are constructed the same way, with two rings, and I timed the cycle at 3¼ seconds, resulting in a considerably more impatient sound than the North American telephone. (Also, on New Zealand telephones no manual adjustment of the intensity of the ringing is possible.) By contrast, in Sweden and parts of Germany the interval between rings is ten seconds. But in 1975 the Swedes began a conversion program to a faster-paced ring. For the telephone company time is money. Lines will be less tied up if telephones are answered more quickly. Thus in order for the telephone company to save a few crowns a whole nation is going to be made more jumpy.

One of the more interesting utilizations of music in a public system was that of the French telephone. When one used to dial 10 (*manuel Paris vers province*), one heard the first measures of Gustave Charpentier's opera *Louise* while the call was being connected; when one dialed 19 (*automatique international*), it was the first bars of the "Ode to Joy" from Beethoven's Ninth Symphony. But on December 4, 1971, the music was replaced by a steady tone of 850 hertz. One of the strategists explains why: "This simplification will contribute to the development of the 'télé-informatique' by permitting the 'télécommande' of all calls under computer control, and besides this will facilitate the installation of punched-key 'numérotation' and the employment of automatic transmitters." Did the public prefer 850 hertz to Beethoven? "The public has not reacted in any noteworthy manner to the recent change, which had already been announced to them well in advance on radio, television, and in the press." The fact was that the music not only slowed down operational efficiency, but it also slowed down telephone users. "Even before the modification we realized that people telephoning in other parts of France, where there was no music, made their calls more quickly than the Parisians, who presumably were lulled by the fine music of *Louise* or the 'Ode to Joy' of Beethoven and paused to listen to it."

The acoustic designer wants to redress some imbalances. He wants to slow things down. He wants to reduce the number of flat-line monotones and to reintroduce strong and exhilarating sounds. You say that he wouldn't have been able to prevent Paris from converting to a monotone, but I say that until he injects himself into a position where he can effectively insist on the aesthetic losses of a one-tone telephone we will continue steering straight toward a social flatland. Who wants to defend Beethoven or Charpentier? Let us have a hundred, a thousand, a million sounds, one for each exchange, for each hamlet, for each customer the world over . . .!

Instead we are told to prepare for international standardization. On the new North American telephones, dialing is speeded up by a punch-number system. Each number on the dial is made up of two frequencies, a low and a high so that tunes are possible, and Beethoven (approximately) returns with the opening of his Fifth as 0005-8883.

Car Horns   Car horns are another example of a sonic absolute bequeathed anonymously to the world by an inventor who took few music lessons. In North American cars the interval of the two horns is set at a major or minor third. The only three-note horns in use are on luxury Cadillacs and Lincoln Continentals, an aristocratic touch which reminds us of the old days when the fattest princes had the biggest orchestras. The horn of the Lincoln Continental is tuned to sound an augmented triad (two superimposed major thirds).

In Turkish cars, horns are tuned to the interval of a major or minor second. While in some cultures this is considered an exceedingly dissonant diad, there are examples in the Balkans, for instance from certain regions of western Bulgaria, of folk singing in which two voices sing together in major or minor seconds, the singers considering this a consonant interval.

In the interest of preserving idiosyncracies in the world soundscape, some thought should be given to using the characteristic intervals and motifs of local musical cultures in tuning environmental signals of all kinds. In Java, for instance, it might be the unique "shortened" fifth that could serve for the car horn, for this interval is, I understand, found in no other culture, though it is basic to the tuning of Gamelan orchestras and is said to derive from the characteristic call of an island bird.

The Utopian Soundscape   There are times for practical repair work and there are times for huge imaginative excursions into utopia. It does not matter whether such dreams are directly realizable; they give elevation to the spirit, nobility to the mind.

One such utopian dream was Charles Ives's Universe Symphony. This was a work with hundreds or thousands of participants, spread out across the valleys, on hillsides and on mountaintops. It was to be so gigantic, so inclusive that no single individual could ever assume mastery or control of it. Anyone who wished to do so could add to it. It was only an idea then, but one which excites our imagination enormously. To imagine ourselves as participants in a Universe Symphony is to give more critical attention to our performance than is the case if we merely consider ourselves to be in a dumpyard. We analyze and criticize the music better; we recognize the soloists, the conductors, the prima donnas; we listen to the talents and faults of each. It is here that the acoustic designer can provide us with portions of the score to perform, as in fact has already been done in environmental compositions by numerous young composers.

Music is the key to the utopian soundscape. By comparison, a study of the sounds in utopian literature is disappointing. On the whole, the suggestions of the futurists have been tepid. About all we learn is that Sir Thomas More anticipated and approved of Moozak, or that Edward Bellamy, in *Looking Backward,* anticipated the radio. The most sustained and broadly conceived soundscape of the future was that of Francis Bacon's *New Atlantis,* where he describes special houses devoted to the study of sounds.

> We have also sound-houses, where we practise and demonstrate all sounds and their generation. We have harmonies which you have not, of quarter-sounds and lesser slides of sounds. Divers instruments of music likewise to you unknown, some sweeter than any you have; together with bells and rings that are dainty and sweet. We represent small sounds as great and deep; likewise great sounds, extenuate and sharp; we make divers tremblings and warblings of sounds, which in their original are entire. We represent and imitate all articulate sounds and letters, and the voices and notes of beasts and birds. We have certain helps, which, set to the ear do further the hearing greatly. We have also divers strange and artificial echoes, reflecting the voice many times, and as it were tossing it; and some that give back the voice louder than it came, some shriller and some deeper; yea, some rendering the voice, differing in the letters or articulate sound from that they receive. We have also means to convey sounds in trunks and pipes, in strange lines and distances.

Listening forward from A.D. 1600, Francis Bacon heard, in his mind's ear, most of the sound inventions of the next three hundred fifty years (recording and editing, modulation and transformation of sound, amplification, broadcasting, telephones, headphones and hearing aids). It was all there in embryo, waiting to be predicted by the thoughtful futurist.

But these instruments are now boring us to death. Now it is our turn to anticipate what lies ahead of our ears and minds. You who would design the future world, listen forward with immense leaps of the imagination and intellect, listen forward fifty, one hundred, a thousand years. What do you hear?

EIGHTEEN

~~~~~~~~~~~~~~~~~~~~~~~~~~~~~~~~~~~~

The Soniferous Garden

A garden is a place where nature is cultivated. It is a humanized treatment of landscape. Trees, fruit, flowers, grass are sculpted organically from the wilderness by art and science. Sometimes the garden is drastically clipped and manicured, as in the harshly classical gardens of Versailles and Vienna; elsewhere man has restricted his touch to assisting certain characteristic features of the landscape to flourish.

A true garden is a feast for all the senses. Here is a description of a medieval garden in Baghdad.

> The gate was arched, and over it were vines with grapes of different colours; the red, like rubies; and the black, like ebony. They entered a bower, and found within it fruits growing in clusters and singly, and the birds were warbling their various notes upon the branches: the nightingale was pouring forth its melodious sounds; and the turtle-dove filled the place with its cooing; and the blackbird, in its singing, resembled a human being; and the ring-dove, a person exhilarated by wine. The fruits upon the trees, comprising every description that was good to eat, had ripened . . . and the place beamed with the charms of spring; the river murmured by while the birds sang, and the wind whistled among the trees; the season was temperate, and the zephyr was languishing.

A park is a public garden into which various community recreations are introduced. Theater, music, athletic events, picnics, all or any of these may be possible in the well-designed park. But parks today are not well designed and that is the problem. In modern cities parks are too often leftover pieces of real estate, belted by what is euphemistically called a parkway, which spits its smell and noise over a site chosen by sight. In older cities, the highways have often been added later, invading once-prized sanctuaries with their filth. This can be seen clearly on isobel maps we made for three Viennese parks: Burggarten, Stadtpark and Belvedere Garten. All are today situated beside busy streets. In none does the ambient level drop below 48 dBA and the average is closer to 55 dBA, which is several decibels above the established Speech Interference Level for normal conversation at four meters.

With good reason then do we insist on the necessity today to throw the emphasis back to the acoustically designed park, or what we might more poetically call the soniferous garden. There is but one principle to guide us in this purpose: always to let nature speak for itself. Water, wind, birds, wood and stone, these are the natural materials which like the trees and shrubs must be organically molded and shaped to bring forth their most characteristic harmonies.

A garden may also be a place of human artifacts such as a bench, a trellis or a swing, but they must harmonize with their natural surroundings, indeed appear to have grown out of them. Thus, if synthetic sounds are introduced into the soniferous garden, they should be sympathetic vibrations of the garden's original notes. There is no place for the wired-in music system, such as I have heard in Ankara's beautifully expansive Gençlik Park, and elsewhere. Nor should other electroacoustic tricks, no matter how clever, find their place here. An American sculptor once demonstrated for me a bridge of high-tension cables, to which he had attached contact microphones and an extensive amplifier and loudspeaker system. Every time a fly landed on one of the cables the sound of a howitzer thundered through the forest. One gasped at the prospect of a crow or a gopher entertaining the whole state of Utah.

Let nature speak with its own authentic voices. That is the grand and simple theme of the acoustic designer. The following notes are reflections on past and prospective solutions to this problem.

The Eloquence of Water   It was in the Italian gardens of the Renaissance and the Baroque era that water was given a special elegance and beauty, for its coolness provided a delicious contrast to the outer glare of the summer heat. The Casino of Pope Pius IV, the Villa Lante, the Villa of Val San Zibio—each has its own romance with water, told in endless fountains, streams, reflecting pools and ingenious water jets, through which the gardens shimmered in fine mists of spray. In the Villa Pliniana a foaming mountain torrent from the Val di Calore pours directly through the central apartment. The old house is saturated with freshness and the bare vaulted rooms reverberate with its joyful sound. But nowhere did water surpass the magnificence it achieved in the gardens of the Villa d'Este at Tivoli, near Rome.

From the Anio, drawn up the hillside at incalculable cost and labour, a thousand rills gush downward, terrace by terrace, channelling the stone rails of the balusters, leaping from step to step, dripping into mossy conchs, flashing in spray from the horns of sea-gods and the jaws of mythical monsters, or forcing themselves in irrepressible over-flow down the ivy-matted banks. The whole length of the second terrace is edged by a deep stone channel, into which the stream drips by countless outlets over a quivering fringe of maidenhair. Every side path or flight of steps is accompanied by its sparkling rill, every niche in the retaining-walls has its water-pouring nymph or gushing urn; the solemn depths of green reverberate with the tumult of innumerable streams.

At the Villa d'Este, water throbs through the garden like the inmost vital principle of the whole. But water was not only used organically in these gardens, it was also used with great artifice and cunning in numerous water sculptures. When John Evelyn visited the Villa d'Este in 1645, these water sculptures were still intact and functioning.

In another garden, is a noble aviary, the birds artificial, and singing till an owl appears, on which they suddenly change their notes. Near this is the fountain of dragons, casting out large streams of water with great noise. In another grotto, called Grotto di Natura, is an hydraulic organ; and below this are divers stews and fish ponds, in one of which is the statue of Neptune in his chariot on a seahorse, in another a Triton; and lastly, a garden of simples.

The invention of water birds appears to date back to Ctesibius of Alexandria (*c.* 250 B.C.). The birds were made to sing by forcing a stream of air into them by water pressure, thus using the same principle as Hero of Alexandria's water organ. In fact, in Hero's *Pneumatics* a very sophisticated version of the singing birds is described in which the birds perform in turns. Here the water, entering a closed vessel, expels the air from it through bronze tubes, placed at staggered levels. These tubes are concealed among the branches of a tree and end in whistles attached to the beaks of artificial birds. Vitruvius also mentions such devices as well as "little figures which drink and move; and other things which flatter the pleasure of the eyes and the use of the ears." That similar devices were plentiful in Italian Baroque gardens is verified by John Evelyn's diary, which contains numerous descriptions of such displays. At Frascati, villa of Cardinal Aldobrandini, Evelyn observed

hydraulic organs, and all sorts of singing birds, moving and chirping by force of the water, with several other pageants and surprising inventions. In the centre of one of these rooms, rises a copper ball that continually dances about three feet above the pavement, by virtue of a wind conveyed secretly to a hole beneath it; with many other devices to wet the unwary spectators, so that one can hardly step without wetting to the skin. In one of these theaters of water, is an Atlas spouting up the stream to a very great height; and another monster makes a terrible roaring with a horn; but, above all, the representation of a storm is most natural, with such fury of rain, wind, and thunder, as one would imagine oneself in some extreme tempest.

There is no doubt that this *théâtre d'eau,* as it is called, often ran to extremes. It is clear, however, that with a little thought the acoustic designer could create great adventures with water. The mere fact that water sounds differently when played on different surfaces and materials could be a subject of rich inventions. Imagine a specially contrived parterre, fashioned out of all kinds of materials—woods, bamboos, metals, scalloped stones, shells—arranged with sounding boxes beneath them, under such a common natural event as a rainstorm. There is a suggestion of the part that different materials and resonating shapes can play under streams of water in Evelyn's description of the famous fountains in the garden of Cardinal Richelieu's villa, at Rueil.

> At the further part of this walk is that plentiful, though artificial cascade, which rolls down a very steep declivity, and over the marble steps and basins, with an astonishing noise and fury; each basin hath a jetto in it, flowing like sheets of transparent glass, especially that which rises over the great shell of lead, from whence it glides silently down a channel through the middle of a spacious gravel walk, terminating in a grotto.... We then saw a large and very rare grotto of shell-work, in the shape of Satyrs, and other wild fancies: in the middle stands a marble table, on which a fountain plays in divers forms of glasses, cups, crosses, fans, crowns, etc. Then the fountaineer represented a shower of rain from the top, met by small jets from below. At going out, two extravagant musketeers shot us with a stream of water from their musket barrels. Before this grotto is a long pool into which ran divers spouts of water from leaden escalop basins.

There is a hint here, perhaps not yet fully developed technically, of a water concert which could become the objective of an exciting collaboration between a sculptor and an acoustic designer.

Around the world there are numerous further water devices which could be utilized for aesthetic effects: the water wheel, for instance, which is most attractive when it revolves asymmetrically so that its tempo is now rapid, now slow. In Bali there is an ingenious irrigation system in which large pieces of bamboo on hinges fill up with water from a stream then flip over, spilling water into the rice fields. As each tips back it produces a hollow tapping sound and one may hear amid the continuous bubbling of water a delicate and irregular recital of taps from fifty or more of these little water mills in operation at once. Alter the lengths of the bamboo tubes and a continuous marimba melody would result.

We must completely revise the thinking of the modern designer for whom a drainpipe is merely runoff for effluvia. Imagine a habitat building with staggered roofs from which the water tumbled into many different kinds of tubes and basins, spurted out of all kinds of gargoyles and spouts, flooded windows, slid down oblique surfaces, and caused all kinds of playful automata to pipe, gurgle, revolve or whistle!

The Spirit of the Wind   The polynoise of water, from which the acoustic designer can draw forth limitless variations, has a pneumatic parallel in the Aeolian harp. Here again man constructs the instrument, but nature plays upon it; and the eerie and even frightening sounds which issue from its strings correspond precisely with what we have already written about the deviousness of the wind. Consider this description by E. T. A. Hoffmann of a large-scale wind harp during a storm.

> I had had the weather harp tightened, which, as you know, is stretched above the large fountain; and the storm, like an accomplished musician, played lustily on it as on a giant harmonica. The chords of this huge organ resounded fearfully through the raging and howling of the hurricane. The powerful tones beat faster and faster, and one might have been listening to a ballet of the Furies in an unusually grand style such as would never be heard within a stage's canvas walls. Well—in half an hour all was over. The moon came out from behind the clouds; the night wind murmured consolingly in the frightened forest and dried the tears on the dark bushes. From time to time the weather harp jingled like a somber, distant bell.

The German enthusiasm for Aeolian harps reached its peak during the Romantic era. They are frequently described in the novels of Jean Paul (1763–1825) as well as in Hoffmann; and Goethe calls for several such instruments to orchestrate the angelic choir in the second part of *Faust* (1832). This association of the Aeolian harp with the soaring spirit of the Faustian temperament illustrates Jung's archetypal coupling of the soul with the transcendental breath of the pneuma.

     The Aeolian harp was claimed as a German invention by Athanasius Kircher in his *Musurgia Universalis* of 1650, where he calls it a "musical autophone," but further research reveals that it was known in Italy, at least in principle, a century before. The invention may be Chinese, by whom it was incorporated into the design of certain kites and was called *Feng Cheng*.

> This is a bow made entirely of bamboo. The string is a very thin slip of bamboo about half an inch wide with a small piece left thick at each end to catch in notches which are cut in the end of the bow; and this is a piece of whole bamboo two or three feet long. It is tied to the frame of a paper kite so that the string will catch the wind just above the head of the kite.

A similar type of wind harp installed in a kite is known in Java. Unlike the European variety, which generally consists of a number of strings, sometimes tuned in a harmonic series, those of Java and China produced single tones, though they still possessed the same unearthly quality. A description of a Javanese wind harp bears this out.

> We heard the deep fundamental of the *sabangan* as F below middle C, which lasted for a long semibreve (say 6 seconds), and then became about half a tone lower. It remained, fading, for a few seconds, and then on the last quarter of the second "bar" passed over into that alarming growling cry, after which the note F was again heard. And so it went on.

The principle of the Aeolian harp was widely known. It is encountered in Ethiopia, South Africa and among the Indians of South America. It is also present in the singing tree of *The Thousand and One Nights,* where beautiful notes arose when even a breeze passed the tree and were then lost upon it. Usually the sound of Aeolian harps produced a peculiar wailing quality which Berlioz likened to "a sharp attack of spleen linked to a temptation toward suicide"; but in fact any number of sounds could be produced by such instruments, and the sculptor and acoustic designer should combine their skills to bring about the rich variety of possible effects for the soniferous garden of the future.

Wind chimes of glass, shell, bamboo, and wood are other means of giving the wind an additional voice, though in this case its sounding is altered to produce a clattering or trembling pulsation of indeterminate character.

A judicious placement of signs in the garden would serve not only to draw the attention of the public to some of its sonic attractions, but also to stimulate that special composure of the mind that the park, of all places in modern society, ought to seek to rejuvenate.

In one corner of the soniferous garden, if it were spacious enough to permit a multiplicity of sonic attractions without becoming a jumble, there might also be a place for a public instrumentarium, such as that conceived by John Grayson. This consists of a number of simple instruments constructed from natural materials, designed to be permanently installed in a park so that the citizens of a community might come together and play together. I would regard this as a most desirable undertaking in the modern world where all activities which tend to reintroduce the feeling of community are valid. Let the Balinese Gamelan orchestra be our model here. In Bali there are no professional musicians; the orchestras are staffed by all the able-bodied of the community, and they strike up in the evening after work and play late into the night.

In the specifications for Grayson's orchestra, the inventor requests an ambient noise level of no more than 45 decibels for his instrumentarium; and the total sound level of all the instruments together is designed not to exceed 80 decibels—that is, it does not exceed the level of the human voice and is accordingly ecologically in balance.

There is no place for the unbalanced sound ecology in the park. The task of the acoustic designer is to find reinforcements of natural sounds in the same way as the trellis reinforces the presence of the rose. One of his special problems will be to return an area of the park to the state of a quiet grove in the midst of active city life. This will not be easy. For frontage on busy streets giant mounds of earth may be the only answer, and these should not only be of sufficient height to hide the traffic from sight, but of such construction as to refract the sound away from the park and of such thickness as to deaden ground vibrations. Sunken gardens, grottoes and other types of acoustic baffles will also be of value here.

The suggestions made in these paragraphs will not be suitable for every park. Above all the acoustic park should be kept simple, and it is for this reason that its chief adornment may be nothing more than the Temple of Silence, a building with no purpose other than meditation. There is nothing special about the Temple of Silence except that in it all visitors will be expected to observe silence. It is to this place that the weary may come seeking nothing but the simplicity of the ultimate music on the other side of this world, the silence at the center of which may be heard the ringing of the great orbs of the Music of the Spheres.

NINETEEN

~~~~~~

Silence

A sound of silence on the startled ear . . .
Edgar Allan Poe, "Al Aaraaf"

Quiet Groves and Times   In the past there were muted sanctuaries where anyone suffering from sound fatigue could go into retirement for recomposure of the psyche. It might be in the woods, or out at sea, or on a snowy mountainside in winter. One would look up at the stars or the soundless soaring of the birds and be at peace.

> Leaning on our stout oaken walking sticks, our sacks on our backs, we climbed the cobbled road that led to Karyés, passing through a dense forest of half-defoliated chestnut-trees, pistachios, and broad-leafed laurels. The air smelled of incense, or so it seemed to us. We felt that we had entered a colossal church composed of sea, mountains and chestnut forests, and roofed at the top by the open sky instead of a dome. I turned to my friend; I wanted to break the silence which had begun to weigh upon me. "Why don't we talk a little?" I suggested. "We are," answered my friend, touching my shoulder lightly. "We are, but with silence, the tongue of angels." Then he suddenly appeared to grow angry. "What do you expect us to say? That it's beautiful, that our hearts have sprouted wings and want to fly away, that we've started along a road leading to Paradise? Words, words, words. Keep quiet!"

Just as man requires time for sleep to refresh and renew his life energies, so too he requires quiet periods to regain mental and spiritual composure. At one time stillness was a precious article in an unwritten code of human rights. Man held reservoirs of stillness in his life to restore the spiritual metabolism. Even in the hearts of cities there were the dark, still vaults of churches and libraries, or the privacy of drawing room and bedroom. Outside the throb of cities, the countryside was accessible with its lulling whirr of natural sounds. There were still times too. The holy days were quieter before they became holidays. In North America, Sunday was the quietest day before it became Fun-day. The importance of these quiet groves and times far transcended the particular purposes to which they were put. We can comprehend this clearly only now that we have lost them.

Ceremonies of Silence   In the park near the Botanic Gardens in Melbourne there is a sign:

IN MEMORY OF
EDWARD GEORGE HONEY

1855–1922

A Melbourne journalist, who,
while
living in London, first suggested
the solemn ceremony of

SILENCE

now observed in all British countries
in remembrance of those who died
in the War.

The fact is that as the memory of the world wars has receded, the observance of silence at 11 a.m. on November 11 has each year become more straggled. It will be the responsibility of the acoustic designers to work not only for the repatriation of quiet groves, but also to lobby for the reintroduction of quiet times. As a matter of fact, Yehudi Menuhin, President of the International Music Council of UNESCO, proposed at the 1975 congress that World Music Day should in the future be celebrated by a minute of silence. We are discussing here something much more important than setting time limits on noisy sounds; we are discussing the deliberate celebration of stillness, which, when observed by an entire society together, is breathtakingly magnificent. Here is an example: the program of the War Remembrance as commemorated each May 4 in Utrecht, Netherlands.

6:00 p.m.   Lowering of flags to half mast in the entire city, until darkness falls. Closing of public amusements. No advertising or store-window lighting.

7:15 p.m.	Participants in the *Silent Procession* will form in threes in St. Peter's churchyard. The places for relatives of the deceased, and other participants will be indicated on signs. People are asked not to carry ensigns, flags or wreaths with them.
7:30–8:00 p.m.	The procession will slowly make its way beneath the sound of all church bells. During the procession, people are requested *to be still* (literally, to pay attention to being silent). The route: St. Peter's churchyard to the Cathedral Square via Voetius Street, Cathedral Street, the Old Church Square, Choir Street, Servet Street and under the Cathedral Tower.
8:00 p.m. precisely	*The bells end and two minutes of total silence begin.* This is indicated by the first of eight chimes of the Cathedral Clock and the lighting of the Cathedral Square.
8:02 p.m.	End of two minutes' silence. The Royal Utrecht PTT Brass Band will play two couplets from the Wilhelmus, sung by those present. During this a wreath will be placed at the foot of the Memorial to the Fallen on behalf of the entire citizenry of Utrecht. All participants in the procession will file past the Memorial and will have the opportunity of laying the flowers brought with them. Everyone is urged to co-operate so that this may be carried out with as much stillness as possible.
8:15–8:45 p.m.	An organ recital in the church by Stoffel van Viegen, closed by the singing of two couplets from the Wilhelmus. Participation is open to everyone.

Attending this ceremony Barry Truax recalled:

It is a unique acoustic ritual in the community. Nothing in the experience of a North American can match it for depth of emotion. As you approach the square, the thundering mass of the largest Cathedral bells rolls over you, enforcing a hypnotic and fearful silence on everyone gathering. The entire weight of the tragedy of the War seems expressed in the heavy low-pitched mass of sound emanating from the high tower.

Slowly, one by one, the bells end and the texture thins as the procession emerges from the passageway under the Tower and slowly divides into rows in front of the Memorial.

The noisy city has become deathly quiet. Now the silence seems as oppressive as the bells did a few moments before. That heavy bombardment seems to have cleansed the air of the city's usual profanity, leaving a strange and nervous calm.

Very quietly a handful of musicians sound the opening chords of the National Anthem in muted low registers. There is an electric moment as a slow unison vibration is born in the throats of all present. The ground itself seems to rise to emit a resonating cry, slowly rising and turning around you in every direction. For a moment the unity these gentle and defiant people felt in the face of the Occupation seems rekindled.

Yet the military is absent. Slowly the individual mourners file past the Memorial to lay their own flowers after the young lad and girl have lifted the city's wreath into place. The number of mourners has fallen off in recent years, but for these few, the experience is relived in a profound and beautiful ceremony, which ends as we enter the Cathedral to the reverberant tones of the organ.

Western Man and Negative Silence   Man likes to make sounds to remind himself that he is not alone. From this point of view total silence is the rejection of the human personality. Man fears the absence of sound as he fears the absence of life. As the ultimate silence is death, it achieves its highest dignity in the memorial service.

Since modern man fears death as none before him, he avoids silence to nourish his fantasy of perpetual life. In Western society, silence is a negative, a vacuum. Silence for Western Man equals communication hang-up. If one has nothing to say, the other will speak; hence the garrulity of modern life which is extended by all kinds of sonic jabberware.

The contemplation of absolute silence has become negative and terrifying for Western Man. Thus when the infinity of space was first suggested by Galileo's telescope, the philosopher Pascal was deeply afraid of the prospect of eternal silence. "Le silence éternel de ces espaces infinis m'effraie."

When one stays for a while in an anechoic chamber—that is, a completely soundproof room—one feels a little of the same terror. One speaks and the sound seems to drop from one's lips to the floor. The ears strain to pick up evidence that there is still life in the world. When John Cage went into such a room, however, he heard two sounds, one high and one low. "When I described them to the engineer in charge, he informed me that the high one was my nervous system in operation, the low one my blood in circulation." Cage's conclusion: "There is no such thing as silence. Something is always happening that makes a sound."

When man regards himself as central in the universe, silence can only be considered as approximate, never absolute. Cage detected this relativity and in choosing *Silence* as the title for his book, he emphasized that for modern man any use of this term must be qualified or assumed to be ironical. Edgar Allan Poe touched on the same thing when in "Al Aaraaf" he wrote: "Quiet we call 'Silence'—which is the merest word of all."

The negative character of silence has made it the most potentialized feature of Western art, where nothingness constitutes the eternal threat to being. Because music represents the ultimate intoxication of life, it is carefully placed in a container of silence. When silence precedes sound, nervous anticipation makes it more vibrant. When it interrupts or follows sound, it reverberates with the tissue of that which sounded, and this state continues as long as the memory holds it. Ergo, however dimly, silence sounds.

Because it is being lost, the composer today is more concerned with silence; he composes with it. Anton Webern moved composition to the brink of silence. The ecstasy of his music is enhanced by his sublime and stunning use of rests, for Webern's is music composed with an eraser. What irony, that the last sound of his life was the explosion of the soldier's gun that shot him.

In *Dummiyah* the Canadian composer John Weinzweig has the conductor conduct long passages of silence in memory of Hitler's victims. "Silence," he says, "is the final sound of the Nazi holocaust."

> In dumb silence I held my peace.
> So my agony was quickened,
> and my heart burned within me.
> [Psalms 39:2–3]

Simultaneously with Webern's discovery of the value of silence in music, his compatriot Freud discovered its value for psychoanalysis. "The analyst is not afraid of silence. As Saussure remarked, the unconnected monologue of the patient on the one side and the almost absolute silence of the psychiatrist on the other was never made a methodological principle before Freud."

The relationship between music and psychoanalysis is by no means fortuitous. Like the music teacher, Freud made regular appointments to see his patients and listened to them at length. In psychoanalysis, as in much modern poetry, that which is not said is pregnant with potential meaning. Philosophy too terminates in silence. Wittgenstein wrote: "Whereof one cannot speak, thereof one must remain silent."

But these things do not reduce my contention that for Western Man silence somehow represents an unutterable impasse, a negative state beyond the realm of the possible, of the attainable. The same semantic complexion is borne out in Western lexicography. The following is the complete entry under "Silence" in *Roget's New Pocket Thesaurus* (New York, 1969). Read it and you will understand that what is described is not a felicitous or positive state but rather merely the muzzling of sound.

> SILENCE—N. silence, quiet, quietude, hush, still; sullenness, sulk, saturninity, taciturnity, laconism, reticence, reserve.
> *muteness,* mutism, deaf-mutism, laloplegia, anarthria, aphasia, aphonia, dysphasia.

speech impediment, stammering, stuttering, baryphony, dysphonia, par-
alalia.

dummy, sphinx, sulk, sulker, calm; mute, deaf-mute, laloplegic,
aphasiac, aphonic, dysphasiac.

    v. *silence,* quiet, quieten, still, hush; gag, muzzle, squelch, tongue-
tie; muffle, stifle, strike dumb.

be silent, quiet down, quiet, hush, dummy up, hold one's tongue, sulk,
say nothing, keep silent, shut up (slang).

    ADJ. *silent,* noiseless, soundless, quiet, hushed, still.

speechless, wordless, voiceless, mute, dumb, inarticulate, tongue-tied,
mousy, mum, sphinxian.

sullen, sulky, glum, saturnine.

taciturn, uncommunicative, close-mouthed, tight-lipped, unvocal,
nonvocal, laconic; reticent, reserved, shy, bashful.

unspoken, tacit, wordless, implied, implicit, understood, unsaid, unut-
tered, unexpressed, unvoiced, unbreathed, unmentioned, untold.

unpronounced, mute, silent, unsounded, surd, voiceless.

inaudible, indistinct, unclear, faint, unheard.

inexpressible, unutterable, indescribable, ineffable, unspeakable,
nameless, unnamable; fabulous.

    See also MODESTY, PEACE. Antonyms—see LOUDNESS, SHOUT.

The Recovery of Positive Silence

In the West we may as-
sume that silence as a condition of life and a workable concept disappeared
sometime toward the end of the thirteenth century, with the death of
Meister Eckhart, Ruysbroeck, Angela de Foligno and the anonymous English author of *The Cloud of Unknowing.* This is the era of the last great
Christian mystics and contemplation as a habit and skill began to disappear
about that time.

Today, as a result of increasing sonic incursions, we are even beginning to lose an understanding of the word concentration. The words sur-
vive all right, that is to say, their skeletons lie in dictionaries; but there are
few who know how to breathe life into them. A recovery of contemplation
would teach us how to regard silence as a positive and felicitous state in
itself, as the great and beautiful backdrop over which our actions are
sketched and without which they would be incomprehensible, indeed
could not even exist. There have been numerous philosophies expressing
this idea and we know that great periods of human history have been
conditioned by them. Such was the message of Lao-tzu: "Give up haste
and activity. Close your mouth. Only then will you comprehend the spirit
of Tâo."

No philosophy or religion catches the positive felicity of stillness
better than Taoism. It is a philosophy that would make all noise abatement
legislation unnecessary. This is also the message of Jalal-ud-din Rumi, who

advised his disciples to "Keep silence like the points of the compass, for the king has erased thy name from the book of speech." Rumi sought to discover that world where "speaking is without letters or sounds." Even today one may observe Bedouins sitting quietly in a circle saying nothing, caught perhaps somewhere between the past and the future—for silence and eternity are bound in mystic union. I recall also the slow stillness of certain Persian villages, where there is still time to sit or squat and think, or merely to sit or squat; time to walk very slowly alongside a child on crutches or a blind grandfather; time to await food or the passage of the sun.

We need to regain quietude in order that fewer sounds can intrude on it with pristine brilliance. The Indian mystic Kirpal Singh expresses this eloquently:

> *The essence of sound* is felt in both motion and silence, it passes from *existent to nonexistent.* When there is no sound, it is said that there is no hearing, but that does not mean that hearing has lost its preparedness. Indeed, when there is no sound, hearing is most alert, and when there is sound the hearing nature is least developed.

When there is no sound, hearing is most alert. It is the same idea that Rilke expresses in his *Duino Elegies* when he speaks of "die ununterbrochene Nachricht, die aus Stille sich bildet." Silence is indeed news for those possessing clairaudience.

If we have a hope of improving the acoustic design of the world, it will be realizable only after the recovery of silence as a positive state in our lives. Still the noise in the mind: that is the first task—then everything else will follow in time.

EPILOGUE

~~~~~~

The Music Beyond

Before man, before the invention of the ear, only the gods heard sounds. Music was then perfect. In both East and West arcane accounts hint at these times. In the *Sangīta-makaranda* (I, 4–6) we learn that there are two forms of sound, the *anāhata,* "unstruck," and the *āhata,* "struck," the first being a vibration of ether, which cannot be perceived by men but is the basis of all manifestation. "It forms permanent numerical patterns which are the basis of the world's existence."

This is identical with the Western concept of the Music of the Spheres, that is, music as rational order, which goes back to the Greeks, particularly to the school of Pythagoras. Having discovered the mathematical correspondence between the ratios of the harmonics in a sounding string, and noting that the planets and stars also appeared to move with perfect regularity, Pythagoras united discovery with intuition and conjectured that the two types of motion were both expressions of a perfect universal law, binding music and mathematics. Pythagoras is reported to have been able to hear the celestial music, though none of his disciples was able to do so. But the intuition persisted. Boethius (A.D. 480–524) also believed in the Music of the Spheres.

> How indeed could the swift mechanism of the sky move silently in its course? And although this sound does not reach our ears (as must for many reasons be the case), the extremely rapid motion of such great bodies could not be altogether without sound, especially since the courses of the stars are joined together by such mutual adaptation that nothing more equally compacted or united could be imagined. For some are borne higher and others lower, and all are revolved with a just impulse, and from their different inequalities an established order

of their courses may be deduced. For this reason an established order of modulation cannot be lacking in this celestial revolution.

If one knew the mass and velocity of a spinning object, it would be possible theoretically to calculate its fundamental pitch. Johannes Kepler, who also believed in a perfect system binding music and astronomy, calculated the following pitches for each of the planets.

In Kepler's notation, the pitches looked like this:

In modern notation, like this:

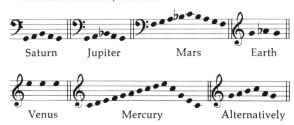

The Music of the Spheres represents eternal perfection. If we do not hear it, it is because we are imperfect. Shakespeare puts it eloquently in *The Merchant of Venice* (V, i).

> Look, how the floor of heaven
> Is thick inlaid with patines of bright gold:
> There's not the smallest orb which thou behold'st
> But in his motion like an angel sings. . . .
> Such harmony is in immortal souls;
> But whilst this muddy vesture of decay
> Doth grossly close it in, we cannot hear it.

But our imperfection is not merely moral; it is physical also. For man, the perfectly pure and mathematically defined sound exists as a theoretical concept only. The French mathematician Fourier knew and stated this when he was developing his theory of harmonic analysis. Distortion results the moment a sound is produced, for the sounding object first has to overcome its own inertia to be set in motion, and in doing this little imperfections creep into the transmitted sound. The same thing is true of our ears. For the ear to begin vibrating, it too has first to overcome its own inertia, and accordingly it too introduces more distortions.

All the sounds we hear are imperfect. For a sound to be totally free of onset distortion, it would have to have been initiated before our lifetime. If it were also continued after our death so that we knew no interruption in it, then we could comprehend it as being perfect. But a sound initiated before our birth, continued unabated and unchanging throughout our lifetime and extended beyond our death, would be perceived by us as —*silence*.

This is why, as I intimated at the beginning of this book, all research into sound must conclude with silence—not the silence of negative vacuum, but the positive silence of perfection and fulfillment. Thus, just as man strives for perfection, all sound aspires to the condition of silence, to the eternal life of the Music of the Spheres.

Can silence be heard? Yes, if we could extend our consciousness outward to the universe and to eternity, we could hear silence. Through the practice of contemplation, little by little, the muscles and the mind relax and the whole body opens out to become an ear. When the Indian yogi attains a state of liberation from the senses, he hears the *anāhata*, the "unstruck" sound. Then perfection is achieved. The secret hieroglyph of the Universe is revealed. Number becomes audible and flows down filling the receiver with tones and light.

Appendixes

APPENDIX I

Sample Sound Notation Systems

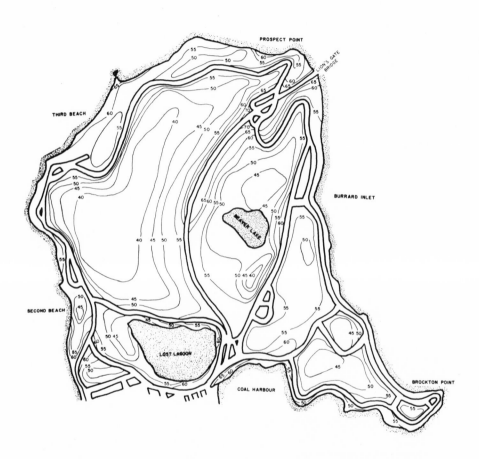

This isobel map of Stanley Park in Vancouver, British Columbia, shows average sound levels in different locations. Sound level measurements were taken on the footpaths at intervals of about 100 yards, between the hours of 10 a.m. and 4 p.m. on several successive Wednesdays during May, June and July of 1973. The weather each day was similar—clear and bright with temperatures in the middle 60s and 70s F. At each point three readings were taken, ten seconds apart, and later averaged together for the construction of the isobels.

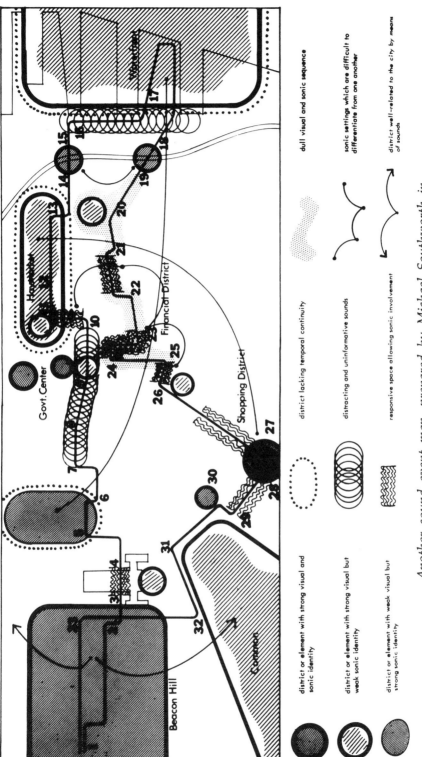

Another sound event map, prepared by Michael Southworth in downtown Boston, that attempts to relate areas with similar and contrasting acoustic environments.

Legend (key terms):

- district or element with strong visual and sonic identity
- district or element with strong visual but weak sonic identity
- district or element with weak visual but strong sonic identity
- district lacking temporal continuity
- distracting and uninformative sounds
- responsive space allowing sonic involvement
- dull visual and sonic sequence
- sonic settings which are difficult to differentiate from one another
- district well-related to the city by means of sounds

Map labels:

Govt. Center
Beacon Hill
Common
Financial District
Shopping District
Waterfront

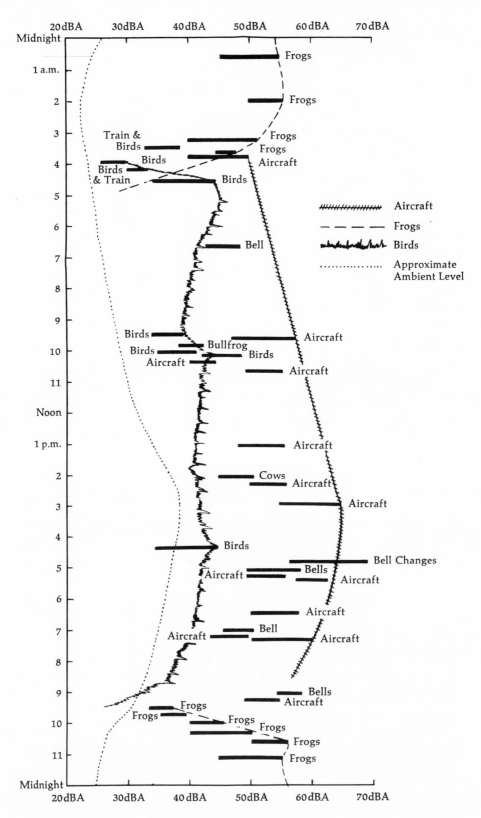

This chart shows log notes of sound events taken during a 24-hour
period in the countryside in British Columbia.

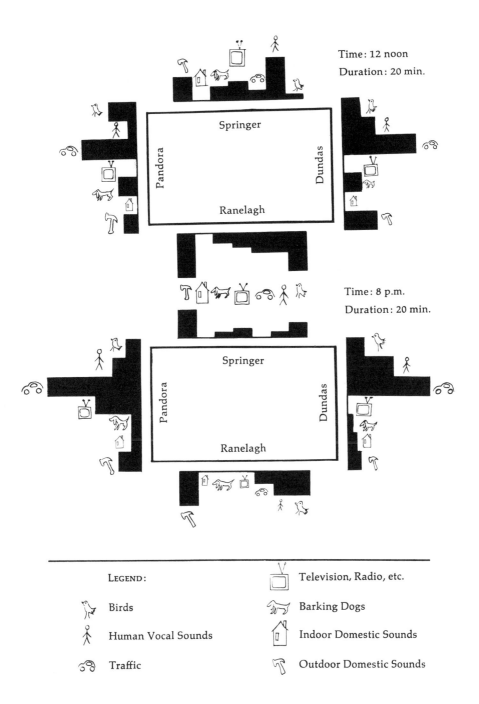

Time: 12 noon
Duration: 20 min.

Springer

Pandora

Dundas

Ranelagh

Time: 8 p.m.
Duration: 20 min.

Springer

Pandora

Dundas

Ranelagh

LEGEND:

Birds

Human Vocal Sounds

Traffic

Television, Radio, etc.

Barking Dogs

Indoor Domestic Sounds

Outdoor Domestic Sounds

One possible form of a sound map, made during two different time periods on a "listening walk" around a city block. The different types of sounds are given graphic values according to whether they are soft, medium or loud, and tabulated to show the general activity and intensity. Using this method, it is simple to make comparisons of sound events historically or geographically.

APPENDIX II

International Sound Preference Survey

Percentage of People Tested Liking or Disliking Sounds by Category

	Auckland, New Zealand 113 People Tested		Vancouver, Canada 99 People Tested		Port Antonio, Jamaica 72 People Tested		Zurich, Switzerland 217 People Tested	
	Pleasant	Unpleasant	Pleasant	Unpleasant	Pleasant	Unpleasant	Pleasant	Unpleasant
WATER								
Rain	31	1	23	0	7	3	25	1
Brooks, Rivers, Waterfalls	18	0	37	0	6	0	43	0
Ocean	58	1	42	0	19	8	4	0
Other	7	0	10	0	0	0	21	2
WIND								
Breeze	50	0	47	0	30	0	28	0
Stormy	0	4	0	0	0	8	1	1
Other	0	0	0	0	0	0	0	0
NATURE								
Dawn	2	0	0	0	0	0	0	0
Night	2	2	0	0	0	7	0	0
Thunderstorms	3	2	2	0	1	6	1	13
Fire Crackling	6	0	8	0	0	0	7	0
Trees	1	1	5	0	0	3	29	1
Other Nature Sounds	1	0	0	0	0	6	7	1
Animals	20	7	22	16	33	100	20	15
Birds	49	3	53	0	68	13	75	7
Insects	10	13	2	5	10	18	15	5

	1	2	3	4	5	6	7	8
HUMAN SOUNDS								
Voices	27	43	35	35	11	60	13	16
Baby Sounds	2	12	2	8	8	11	0	4
Laughter	27	3	20	2	31	6	6	0
Crying	10	16	0	23	0	40	0	7
Body (Breathing, Belching, Snoring, etc.)	8	9	13	21	7	15	2	6
Whistling	1	0	2	0	17	0	0	2
Lovemaking	6	0	8	0	0	0	0	0
Footsteps	3	4	3	0	0	3	3	4
Other	1	3	3	3	1	14	1	11
MUSIC								
Specific Instruments	29	0	35	0	58	0	29	4
Vocal	23	0	12	0	49	0	7	4
Types of Music (Jazz, Classical)	13	4	4	17	15	0	9	1
Other Mentions	28	10	17	3	35	7	40	1
SOUND EQUIPMENT								
Amplifiers	0	0	0	6	0	1	0	1
Malfunctioning Equipment	0	0	0	8	0	0	0	1
Radio and T.V. Commercials	0	9	0	7	0	0	0	0
Other	0	0	0	2	4	0	4	1
DOMESTIC								
Door Slam	0	10	4	0	0	8	0	12
Clocks	2	12	1	6	0	0	4	8
Telephone	2	6	0	5	0	1	1	13
Other	9	4	10	19	1	18	5	14
TRANSPORTATION								
Traffic Noise	0	43	0	32	0	0	4	6
Specific Vehicles Mentioned	8	30	6	58	13	26	4	94

	AUCKLAND, NEW ZEALAND 113 People Tested		VANCOUVER, CANADA 99 People Tested		PORT ANTONIO, JAMAICA 72 People Tested		ZURICH, SWITZERLAND 217 People Tested	
	Pleasant	Unpleasant	Pleasant	Unpleasant	Pleasant	Unpleasant	Pleasant	Unpleasant
TRANSPORTATION (continued)								
Aircraft	1	4	0	5	7	0	2	36
Trains	0	1	3	1	1	0	4	6
Sounds of Accidents	0	6	0	1	0	4	0	1
MACHINERY AND MECHANICAL								
Machinery (General)	0	23	1	19	0	0	2	46
Construction	0	11	0	10	0	0	0	15
Jackhammers	0	15	0	13	0	0	0	14
Dentist Drills	0	12	0	13	0	0	0	5
Power Lawnmowers	0	18	1	0	0	0	0	3
Sirens	0	15	0	25	0	0	0	26
Other	1	12	0	27	0	0	0	18
OTHER SOUNDS								
Bells	2	0	8	0	1	0	54	2
Loud Impact (Gunshot, etc.)	0	8	0	7	1	4	1	13
Hammering	0	4	0	7	0	0	0	1
Chalk Squeaking on Blackboard	0	38	0	32	0	1	0	13
Miscellaneous	4	8	11	1	1	4	2	2
Silence	8	0	15	0	0	0	1	1

Glossary of Soundscape Terms

The following short list of terms includes only neologisms or acoustic terms which I have adapted and given special meanings to for the purpose of this book. The list does not include general acoustic terms employed in the customary manner, definitions of which may be found in standard works of reference.

ACOUSTIC DESIGN: A new interdiscipline requiring the talents of scientists, social scientists and artists (particularly musicians), acoustic design attempts to discover principles by which the aesthetic quality of the acoustic environment or SOUNDSCAPE may be improved. In order to do this it is necessary to conceive of the soundscape as a huge musical composition, ceaselessly evolving about us, and to ask how its orchestration and forms may be improved to bring about a richness and diversity of effects which, nevertheless, should never be destructive of human health or welfare. The principles of acoustic design may thus include the elimination or restriction of certain sounds (noise abatement), the testing of new sounds before they are released indiscriminately into the environment, but also the preservation of sounds (SOUNDMARKS), and above all the imaginative placement of sounds to create attractive and stimulating acoustic environments for the future. Acoustic design may also include the composition of model environments, and in this respect it is contiguous with contemporary musical composition. Compare: ACOUSTIC ECOLOGY.

ACOUSTIC ECOLOGY: Ecology is the study of the relationship between living organisms and their environment. Acoustic ecology is thus the study of the effects of the acoustic environment or SOUNDSCAPE on the physical responses or behavioral characteristics of creatures living within it. Its particular aim is to draw attention to imbalances which may have unhealthy or inimical effects. Compare: ACOUSTIC DESIGN.

ACOUSTIC SPACE: The profile of a sound over the landscape. The acoustic space of any sound is that area over which it may be heard before it drops below the ambient sound level.

AURAL SPACE: The space on any graph which results from a plotting of the various dimensions of sound against one another. For convenience in reading usually only two dimensions are plotted at once. Thus time may be plotted against frequency, frequency against amplitude or amplitude against time. Aural space is thus merely a notational convention and should not be confused with ACOUSTIC SPACE, which is an expression of the profile of a sound over the landscape.

CLAIRAUDIENCE: Literally, clear hearing. The way I use the term there is nothing mystical about it; it simply refers to exceptional hearing ability, particularly with regard to environmental sound. Hearing ability may be trained to the clairaudient state by means of EAR CLEANING exercises.

EAR CLEANING: A systematic program for training the ears to listen more discriminatingly to sounds, particularly those of the environment. A set of such exercises is given in my book *Ear Cleaning.*

EARWITNESS: One who testifies or can testify to what he or she has heard.

HI-FI: Abbreviation for high fidelity, that is, a favorable signal-to-noise ratio. The most general use of the term is in electroacoustics. Applied to soundscape studies a hi-fi environment is one in which sounds may be heard clearly without crowding or masking. Compare: LO-FI.

KEYNOTE SOUND: In music, keynote identifies the key or tonality of a particular composition. It provides the fundamental tone around which the composition may modulate but from which other tonalities take on a special relationship. In soundscape studies, keynote sounds are those which are heard by a particular society continuously or frequently enough to form a background against which other sounds are perceived. Examples might be the sound of the sea for a maritime community or the sound of the internal combustion engine in the modern city. Often keynote sounds are not consciously perceived, but they act as conditioning agents in the perception of other sound signals. They have accordingly been likened to the ground in the figure-ground grouping of visual perception. Compare: SOUND SIGNAL.

LO-FI: Abbreviation for low fidelity, that is, an unfavorable signal-to-noise ratio. Applied to soundscape studies a lo-fi environment is one in which signals are overcrowded, resulting in masking or lack of clarity. Compare: HI-FI.

MOOZAK (MOOZE, etc.): Term applying to all kinds of schizophonic musical drool, especially in public places. Not to be confused with the brand product Muzak.

MORPHOLOGY: The study of forms and structures. Originally employed in biology, it was later (by 1869) employed in philology to refer to patterns of inflection and word formation. Applied to soundscape studies it refers to changes in groups of sounds with similar forms or functions when arbitrarily arranged in temporal or spatial formations. Examples of acoustic morphology might be a study of the historical evolution of foghorns, or a geographical comparison of methods of telegraphy (alphorn, jungle drums, etc.).

NOISE: Etymologically the word can be traced back to Old French (*noyse*) and to eleventh-century Provençal (*noysa, nosa, nausa*), but its origin is uncertain. It has a variety of meanings and shadings of meaning, the most important of which are the following:

1. *Unwanted sound.* The Oxford English Dictionary contains references to noise as unwanted sound dating back as far as 1225.
2. *Unmusical sound.* The nineteenth-century physicist Hermann Helmholtz employed the expression noise to describe sound composed of nonperiodic vibrations (the rustling of leaves), by comparison with musical sounds, which consist of periodic vibrations. Noise is still used in this sense in expressions such as white noise or Gaussian noise.
3. *Any loud sound.* In general usage today, noise often refers to particularly loud sounds. In this sense a noise abatement by-law prohibits certain loud sounds or establishes their permissible limits in decibels.
4. *Disturbance in any signaling system.* In electronics and engineering, noise refers to any disturbances which do not represent part of the signal, such as static on a telephone or snow on a television screen.

The most satisfactory definition of noise for general usage is still "unwanted sound." This makes noise a subjective term. One man's music may be another man's noise. But it holds open the possibility that in a given society there will be more agreement than disagreement as to which sounds constitute unwanted interruptions. It should be noted that each language preserves unique nuances of meaning for words representing noise. Thus in French one speaks of the *bruit* of a jet but also the *bruit* of the birds or the *bruit* of the waves. Compare: SACRED NOISE.

SACRED NOISE: Any prodigious sound (noise) which is exempt from social proscription. Originally Sacred Noise referred to natural phenomena such as thunder, volcanic eruptions, storms, etc., as these were believed to represent divine combats or divine displeasure with man. By analogy the expression may be extended to social noises which, at least during certain periods, have escaped the attention of noise abatement legislators, e.g., church bells, industrial noise, amplified pop music, etc.

SCHIZOPHONIA (Greek: *schizo* = split and *phone* = voice, sound): I first employed this term in *The New Soundscape* to refer to the split between an original sound and its electroacoustic reproduction. Original sounds are tied to the mechanisms that produce them. Electroacoustically reproduced sounds are copies and they may be restated at other times or places. I employ this "nervous" word in order to dramatize the aberrational effect of this twentieth-century development.

SONIFEROUS GARDEN: A garden, and by analogy any place, of acoustic delights. This may be a natural soundscape, or one submitted to the

principles of ACOUSTIC DESIGN. The soniferous garden may also include as one of its principal attractions a Temple of Silence for meditation.

SONOGRAPHY: The art of soundscape notation. It may include customary
methods of notation such as the sonogram or sound level recording,
but beyond these it will also seek to register the geographic distribution of SOUND EVENTS. Various techniques of aerial sonography are
employed, for instance, the isobel contour map.

SONOLOGICAL COMPETENCE: The implicit knowledge which permits the
comprehension of sound formations. The term has been borrowed
from Otto Laske. Sonological competence unites impression with cognition and makes it possible to formulate and express sonic perceptions. It is possible that just as sonological competence varies from
individual to individual, it may also vary from culture to culture, or
at least may be developed differently in different cultures. Sonological
competence may be assisted by EAR CLEANING exercises. See O. Laske,
"Musical Acoustics (Sonology): A Questionable Science Reconsid-
ered," *Numus-West,* Seattle, No. 6, 1974; "Toward a Theory of Musical Cognition," *Interface,* Amsterdam, Vol. 4, No. 2, Winter, 1975, *inter
alia.*

SOUND EVENT: Dictionary definition of *event:* "something that occurs in a
certain place during a particular interval of time." This suggests that
the event is not abstractable from the time-and-space continuum
which give it its definition. The sound event, like the SOUND OBJECT,
is defined by the human ear as the smallest self-contained particle of
a SOUNDSCAPE. It differs from the sound object in that the latter is an
abstract acoustical object for study, while the sound event is a symbolic, semantic or structural object for study, and is therefore a nonab-
stractable point of reference, related to a whole of greater magnitude
than itself.

SOUNDMARK: The term is derived from *landmark* to refer to a community
sound which is unique or possesses qualities which make it specially
regarded or noticed by the people in that community.

SOUND OBJECT: Pierre Schaeffer, the inventor of this term (*l'objet sonore*),
describes it as an acoustical "object for human perception and not a
mathematical or electro-acoustical object for synthesis." The sound
object is then defined by the human ear as the smallest self-contained
particle of a SOUNDSCAPE, and is analyzable by the characteristics of
its envelope. Though the sound object may be referential (i.e., a bell,
a drum, etc.), it is to be considered primarily as a phenomenological
sound formation, independently of its referential qualities as a sound
event. Compare: SOUND EVENT.

SOUNDSCAPE: The sonic environment. Technically, any portion of the
sonic environment regarded as a field for study. The term may refer
to actual environments, or to abstract constructions such as musical

compositions and tape montages, particularly when considered as an environment.

SOUND SIGNAL: Any sound to which the attention is particularly directed. In soundscape studies sound signals are contrasted by KEYNOTE SOUNDS, in much the same way as figure and ground are contrasted in visual perception.

WORLD SOUNDSCAPE PROJECT: A project headquartered at the Sonic Research Studio of the Communications Department, Simon Fraser University, British Columbia, Canada, devoted to the comparative study of the world SOUNDSCAPE. The Project came into existence in 1971, and since that time a number of national and international research studies have been conducted, dealing with aural perception, sound symbolism, noise pollution, etc., all of which have attempted to unite the arts and sciences of sound studies in preparation for the development of the interdiscipline of ACOUSTIC DESIGN. Publications of the World Soundscape Project have included: *The Book of Noise, The Music of the Environment, A Survey of Community Noise By-Laws in Canada* (1972), *The Vancouver Soundscape, A Dictionary of Acoustic Ecology, Five Village Soundscapes* and *A European Sound Diary*.

Notes

INTRODUCTION

p. 5
Music is sounds   Quoted from R. Murray Schafer, *The New Soundscape,* London and Vienna, 1971, p. 1.

p. 7
Therefore the music   Hermann Hesse, *The Glass Bead Game,* New York, 1969, p. 30.

p. 8
When a writer   Erich Maria Remarque, *All Quiet on the Western Front,* Boston, 1929, see Chapter 4.

p. 9
The days are hot   Ibid., p. 126.

William Faulkner   William Faulkner, *As I Lay Dying,* New York, 1960, p. 202.

p. 11
rural Africans   J. C. Carothers, "Culture, Psychiatry and the Written Word," *Psychiatry,* November, 1959, pp. 308–310.

Terror   Marshall McLuhan, *The Gutenberg Galaxy,* Toronto, 1962, p. 32.

CHAPTER ONE: *The Natural Soundscape*

p. 15
Some say that   Robert Graves, *The Greek Myths* (according to Hera's statement in the *Iliad,* XIV), New York, 1955, p. 30.

the waters little   The Questions of King Milinda, trans. T. W. Rhys Davids, Vol. XXXV of *The Sacred Books of the East,* Oxford, 1890, p. 175.

p. 16
And poor old Homer   The Cantos of Ezra Pound, London, 1954, p. 10.

For fifty days   Hesiod, *Works and Days,* lines 663–665, trans. R. Lattimore, Ann Arbor, Michigan, 1968.

p. 17

waves roared   *The Saga of the Volsungs,* ed. R. G. Finch, London, 1965, p. 15.

Splashing oars   *First Lay of Helgi,* lines 104–110, trans. by the author.

Waves coming   Laragia tribe, Australia, *Technicians of the Sacred,* ed. J. Rothenberg, New York, 1969, p. 314.

Lithe turning   *The Cantos of Ezra Pound, op. cit.,* pp. 13–14.

p. 18

The wanderer   Thomas Hardy, *The Mayor of Casterbridge,* London, 1920, p. 341.

whirling and sucking   Henry David Thoreau, *A Week on the Concord and Merrimack Rivers,* in *Walden and Other Writings,* New York, 1937, p. 413.

producing a hollow   J. Fenimore Cooper, *The Pathfinder,* New York, 1863, p. 115.

p. 19

For the noise   Emil Ludwig, *The Nile,* trans. M. H. Lindsay, New York, 1937, pp. 250–251.

the soft splash   Somerset Maugham, *The Gentleman in the Parlour,* London, 1940, p. 159.

Water slapped   Thomas Mann, "Death in Venice," *Stories of Three Decades,* New York, 1936, p. 421.

The rain drops   Emily Carr, *Hundreds and Thousands,* Toronto/Vancouver, 1966, p. 305.

the thunder boomed   Alan Paton, *Cry, the Beloved Country,* New York, 1950, p. 244.

p. 20

The Illustrated Glossary   T. Armstrong, B. Roberts and C. Swithinbank, *The Illustrated Glossary of Snow and Ice,* Cambridge, 1966.

In wintertime   George Green, *History of Burnaby and Vicinity,* Vancouver, 1947, p. 3.

p. 21

we glided along   F. Philip Grove, *Over Prairie Trails,* Toronto, 1922, p. 91.

Nor is anything   Hugh MacLennan, *The Watch That Ends the Night,* Toronto, 1961, p. 5.

The violent Russian   Igor Stravinsky, *Memories and Commentaries,* London, 1960, p. 30.

and inside each   Hesiod, *Theogony,* lines 829–835, trans. R. Lattimore, Ann Arbor, Michigan, 1968.

p. 22

Le vaste trouble   Victor Hugo, *Les Travailleurs de la Mer,* Paris, 1869, pp. 191–192.

The wind could   W. O. Mitchell, *Who Has Seen the Wind?,* Toronto, 1947, pp. 191 and 235.

To dwellers   Thomas Hardy, *Under the Greenwood Tree,* London, 1903, p. 3.

p. 23
The silence Emily Carr, *The Book of Small,* Toronto, 1942, p. 119.

p. 24
for, though the quiet J. Fenimore Cooper, *op. cit.,* pp. 104–105.

It is rather difficult Pseudo-Plutarch, *Treatise on Rivers and Mountains.* Quoted from F. D. Adams, *The Birth and Development of the Geological Sciences,* New York, 1954, p. 31.

p. 25
the earth tremors *The Saga of the Volsungs, op. cit.,* pp. 30–31.

his hair stood *The Lay of Thrym,* from *The Elder Edda,* trans. Patricia Terry, New York, 1969, p. 88.

the infinite great Hesiod, *Theogony, op. cit.,* lines 678–694.

Then the Earth Dion Cassius, quoted from Thomas Burnet, *The Sacred Theory of the Earth,* Book III, Chapter VII (1691), Carbondale, Illinois, 1965, p. 275.

p. 26
At the crater Thorkell Sigurbjörnsson, personal communication.

Within three or four David Simmons, personal communication.

I did not reach Heinrich Heine, "Die Harzreise," *Sämtliche Werke,* Vol. 2, Munich, 1969, pp. 19–20.

p. 27
Howl ye Isaiah 13:6 and 13.

p. 28
By the din Jalal-ud-din Rumi, *Divan i Shams i Tabriz.*

They put their fingers Qur'an, 2:19.

Several times *The Eruption of Krakatoa,* Report of the Krakatoa Committee of the Royal Society, London, 1888, pp. 79–80.

CHAPTER TWO: *The Sounds of Life*

p. 29
is so intense A. J. Marshall, "The Function of Vocal Mimicry in Birds," *Emu,* Melbourne, Vol. 50, 1950, p. 9.

p. 30
Hawfinch From E. M. Nicholson and Ludwig Koch, *Songs of Wild Birds,* London, 1946.

p. 31
The quincunxes Victor Hugo, *Les Misérables,* 1862. Quoted from *Landscape Painting of the Nineteenth Century,* Marco Valsecchi, New York, 1971, p. 106.

p. 32
absolutely nothing Ferdinand Kürnberger, *Der Amerika-müde,* 1855. Quoted from David Lowenthal, "The American Scene," *The Geographical Review,* Vol. LVIII, No. 1, 1968, p. 71.

Everything   Nicolai Gogol, *Evenings on a Farm near Dikanka,* 1831–32. Quoted from Marco Valsecchi, *op. cit.,* p. 279.

How enchanting   Boris Pasternak, *Doctor Zhivago,* New York, 1958, p. 11.

What could be   Maxim Gorky, *Childhood.* Quoted from Marco Valsecchi, *op. cit.,* p. 279.

p. 33
The noise   Somerset Maugham, *The Gentleman in the Parlour,* London, 1940, p. 138.

The owl's   F. Philip Grove, *Over Prairie Trails,* Toronto, 1922, p. 35.

p. 34
His ears   Leo Tolstoy, *Anna Karenina,* trans. C. Garnett, New York, 1965, p. 837.

The flight   Leo Tolstoy, *War and Peace,* trans. C. Garnett, London, 1971, p. 944.

a cry   Virgil, *Georgics,* Book IV, lines 62–64 and 70–72, trans. C. Day Lewis, New York, 1964.

p. 35
I remember   Julian Huxley and Ludwig Koch, *Animal Language,* New York, 1964, p. 24.

The bleating   Compare Virgil's *Eclogue II* with Pope's paraphrase, *The Second Pastoral.*

p. 36
May the fallows   Theocritus, *Idyll XVI,* edited and translated by A. S. F. Gow, Vol. 1, Cambridge, 1950, p. 129.

It is not our purpose   A good general survey of this subject, and a book from which we have drawn numerous facts, is *Animal Language* by Julian Huxley and Ludwig Koch, New York, 1964.

p. 37
refused to believe   Julian Huxley and Ludwig Koch, *op. cit.,* p. 41.

p. 40
One must have heard   Marius Schneider, "Primitive Music," *The New Oxford History of Music,* Vol. 1, London, 1957, p. 9.

Now, it is   Otto Jespersen, *Language: Its Nature, Development and Origin,* London, 1964, pp. 420 and 437.

CHAPTER THREE: *The Rural Soundscape*

p. 43
He was disturbed   Thomas Hardy, *Far from the Madding Crowd,* London, 1920, p. 291.

p. 44
When I hear   Johann Wolfgang von Goethe, *Die Leiden des Jungen Werthers* (*The Sorrows of Young Werther*), in *Werke,* Vol. 19, Weimar, 1899, p. 8.

Hyblaean bees   Virgil, *The Pastoral Poems, Eclogue I,* trans. E. V. Rieu, Harmondsworth, Middlesex, 1949.

Sweet   Theocritus, edited and translated by A. S. F. Gow, Vol. I, *Idyll I,* Cambridge, 1950.

p. 45

The music struck   Virgil, *The Pastoral Poems, op. cit., Eclogue VI.*

Practice country   *Ibid., Eclogue I.*

Now and then   Alain-Fournier, *The Wanderer (Le Grand Meaulnes),* trans. L. Bair, New York, 1971, p. 29.

The shepherd   Thomas Hardy, "Fellow Townsmen," *Wessex Tales,* London, 1920, p. 111.

Sigmund blew   *The Saga of the Volsungs,* ed. R. G. Finch, London, 1965, p. 20.

p. 46

The hounds'   Leo Tolstoy, *War and Peace,* trans. C. Garnett, London, 1971, p. 536.

It was still quite dark   Hildegard Westerkamp, personal communication.

p. 47

In Germany   Private communication from the Deutsches Bundesministerium für das Post-und Fernmeldewesen.

In Austria   Dr. Ernst Popp, personal communication.

Through the narrow   Karl Thieme, "Zur Geschichte des Posthorns," in *Posthornschule und Posthorn-Taschenliederbuch,* Friedrich Gumbert, Leipzig, 1908, pp. 6–7.

p. 48

One farmer   Virgil, *Georgics,* Book I, lines 291–296, trans. Smith Palmer Bovie, Chicago, 1956.

p. 49

The grass   Leo Tolstoy, *Anna Karenina,* trans. C. Garnett, New York, 1965, p. 270.

The peasant   *Ibid.,* p. 291.

Such was the life   Virgil, *Georgics,* Book II, lines 538–540, trans. C. Day Lewis, New York, 1964.

At the shouts   *The Epic of the Kings* (Shāh-nāma), trans. Reuben Levy, Chicago, 1967, p. 57.

p. 50

One should send   Onasander, *The General,* XXIX, trans. William A. Oldfather *et al.,* London, 1923, p. 471.

By the rendering   Tacitus, *Germania,* trans. H. Mattingly and S. A. Handford, Harmondsworth, Middlesex, 1970, p. 103.

It was at three   H. G. Wells, *The Outline of History,* New York, 1920, p. 591.

p. 51

In general   Samuel Rosen *et al.,* "Presbycusis Study of a Relatively Noise-Free Population in the Sudan," American Otological Society, *Transactions,* Vol. 50, 1962, pp. 140–141.

CHAPTER FOUR: *From Town to City*

p. 54

One sound rose   Johan Huizinga, *The Waning of the Middle Ages,* New York, 1954, pp. 10–11.

The great convenience   Dr. Charles Burney, *An Eighteenth-Century Musical Tour in Central Europe and the Netherlands,* Vol. II, London, 1959, p. 6.

On the other side   Robert Louis Stevenson, *An Inland Voyage,* New York, 1911, p. 211.

p. 55

The church clock   Thomas Hardy, *Far from the Madding Crowd,* London, 1922, p. 238.

Other clocks   Thomas Hardy, *The Mayor of Casterbridge,* London, 1920, pp. 32–33.

p. 56

Post office   By-law No. 98–63 (1963).

Amongst the Western   Oswald Spengler, *Der Untergang des Abendlandes,* Vol. 1, Munich, 1923, p. 8.

p. 57

Where the lake   Ippolito Nievo, *Confessions of an Octogenarian,* 1867. Quoted from *Landscape Painting of the Nineteenth Century,* Marco Valsecchi, New York, 1971, p. 184.

a remote resemblance   Thomas Hardy, *The Trumpet-Major,* London, 1920, p. 2.

Awakening   Maxim Gorky, *The Artamonovs,* Moscow, 1952, p. 404.

the sounds   W. O. Mitchell, *Who Has Seen the Wind?,* Toronto, 1947, p. 230.

p. 58

We started   James Morier, *The Adventures of Hajji Baba of Ispahan,* New York, 1954, p. 19.

p. 59

The first streets   Eric Nicol, *Vancouver,* Toronto, 1970, p. 54.

p. 60

The curfew   Thomas Hardy, *The Mayor of Casterbridge, op. cit.,* p. 32.

I had   James Morier, *op. cit.,* p. 123.

p. 61

Later that night   Alain-Fournier, *The Wanderer (Le Grand Meaulnes),* trans. L. Bair, New York, 1971, pp. 124–125.

One was a Dandy   Leigh Hunt, *Essays and Sketches,* London, 1912, pp. 73–74.

There was the faint   Virginia Woolf, *Orlando,* London, 1960, p. 203.

p. 62

I go to bed   Tobias Smollett, *The Expedition of Humphry Clinker,* New York, 1966, pp. 136–137.

The creaking   Charles Mair, 1868. Quoted from *Life at Red River,* Keith Wilson, Toronto, 1970, p. 12.

I denounce   Arthur Schopenhauer, "On Noise," *The Pessimist's Handbook,* trans. T. Bailey Saunders, Lincoln, Nebraska, 1964, pp. 217–218.

p. 63

the moist circuit   Leigh Hunt, *op. cit.,* p. 258.

Pyotr Artamonov and *gather on the bank*   *Op. cit.,* pp. 22–23.

p. 64

Labor   Lewis Mumford, *Technics and Civilization,* New York, 1934, p. 201.

Turn your eyes   Renato Fucini, *Naples Through a Naked Eye,* 1878. Quoted from Marco Valsecchi, *op. cit.,* p. 182.

Usually   Johann Friedrich Reichardt, *Vertraute Briefe aus Paris Geschrieben in den Jahren 1802 und 1803,* Erster Theil, Hamburg, 1804, p. 252.

p. 65

13 different   Sir Frederick Bridge, "The Musical Cries of London in Shakespeare's Time," *Proceedings of the Royal Musical Association,* Vol. XLVI, London, 1919, pp. 13–20.

Between the acts   Johann Friedrich Reichardt, *op. cit.,* pp. 248–249.

p. 66

Your correspondents   Michael T. Bass, *Street Music in the Metropolis,* London, 1864, p. 41.

one-fourth part   Charles Babbage, *Passages from the Life of a Philosopher,* London, 1864, p. 345.

CHAPTER FIVE: *The Industrial Revolution*

p. 73

A hasty lunch   Thomas Hardy, *Tess of the d'Urbervilles,* Vol. 1, London, 1920, p. 416.

The little   Stendhal, *Le Rouge et le Noir,* Paris, 1927, pp. 3–4.

p. 74

A vague   Edmond and Jules de Goncourt, *Renée Mauperin,* 1864. Quoted from *Landscape Painting of the Nineteenth Century,* Marco Valsecchi, New York, 1971, p. 107.

We are encompassed   Thomas Mann, "A Man and His Dog," *Stories of Three Decades,* New York, 1936, pp. 440–441.

As they worked   D. H. Lawrence, *The Rainbow,* New York, 1943, p. 6.

p. 75

I happened   Report of the Sadler Factory Investigating Committee, London, 1832, p. 99.

Is one part   *Ibid.,* p. 159.

Stephen bent Charles Dickens, *Hard Times for These Times,* London, 1955, p. 69.

And now it had occurred Émile Zola, *Germinal,* Harmondsworth, Middlesex, 1954, p. 311.

Fosbroke Cf. John Fosbroke, "Practical Observations on the Pathology and Treatment of Deafness," *Lancet,* VI, 1831, pp. 645–648; and T. Barr, "Enquiry into the Effects of Loud Sounds upon Boilermakers," *Proceedings of the Glasgow Philosophical Society,* 17, 1886, p. 223.

p. 80
musically Jess J. Josephs, *The Physics of Musical Sounds,* Princeton, N.J., 1967, p. 20.

Night and day Charles Dickens, *Dombey and Son,* London, 1950, p. 219.

p. 81
Louder *Ibid.,* p. 281.

Then the shrill D. H. Lawrence, *op. cit.,* p. 6.

The Canadian Jean Reed, personal communication.

p. 83
the trend David Apps, General Motors; personal communication.

p. 84
present a definite *Snowmobile Noise, Its Sources, Hazards and Control,* APS–477, National Research Council, Ottawa, 1970.

p. 85
Air travel E. J. Richards, "Noise and the Design of Airports," *Conference on World Airports—The Way Ahead,* London, 1969, p. 63.

p. 86
a growth *Ibid.,* p. 69.

On the basis William A. Shurcliff, *SST and Sonic Boom Handbook,* New York, 1970, p. 24.

CHAPTER SIX: *The Electric Revolution*

p. 91
We should not "Ohne Kraftwagen, ohne Flugzeug und ohne Lautsprecher hätten wir Deutschland nicht erobert," Adolf Hitler, *Manual of the German Radio,* 1938–39.

When I go Emily Carr, *Hundreds and Thousands,* Toronto/Vancouver, 1966, pp. 230–231.

p. 92
At once Hermann Hesse, *Steppenwolf,* New York, 1963, p. 239.

It takes hold *Ibid.,* p. 240.

p. 94
in the fact that Sergei M. Eisenstein, *The Film Sense,* trans. Jay Leyda, London, 1943, p. 14.

p. 97

MUZAK   Classified Section, *Vancouver Telephone Directory,* British Columbia Telephone Company, 1972, p. 424.

program specialists   Environs, Vol. 2, No. 3, published by the Muzak Corporation.

Each 15-minute   Ibid.

p. 98

Music Library   Memo from the firm of Bolt Berenak and Newman to Dr. Robert Fink, Head, Music Department, Western Michigan University.

p. 99

The modal group   Alain Daniélou, *The Raga-s of Northern Indian Music,* London, 1968, pp. 22–23.

CHAPTER SEVEN: *Music, the Soundscape and Changing Perceptions*

p. 104

easy and insatiable   The phrase is Raymond Williams's.

urban activity   See *The Oxford Companion to Music,* London, 1950, p. 900.

p. 105

At their due times   Gottfried von Strassburg, *Tristan,* trans. A. T. Hatto, Harmondsworth, Middlesex, 1960, pp. 262–263.

p. 108

with the increase   Lewis Mumford, *Technics and Civilization,* New York, 1934, pp. 202–203.

p. 109

Complaints   Quoted from Kurt Blaukopf, *Hexenküche der Musik,* Teufen, Switzerland, 1959, p. 45.

p. 110

signifies   Oswald Spengler, *Der Untergang des Abendlandes,* Munich, 1923, Vol. I, p. 375.

I take it that music   Ezra Pound, *Antheil and the Treatise on Harmony,* New York, 1968, p. 53.

In antiquity   Luigi Russolo, *The Art of Noises,* New York, pp. 3–8.

p. 116

It is recorded   Michel P. Philippot, "Observations on Sound Volume and Music Listening," *New Patterns of Musical Behaviour,* Vienna, 1974, p. 55.

p. 118

The sound   Kurt Blaukopf, "Problems of Architectural Acoustics in Musical Sociology," *Gravesaner Blätter,* Vol. V, Nos. 19/20, 1960, p. 180.

CHAPTER EIGHT: *Notation*

p. 127

To render   Hermann Helmholtz, *On the Sensations of Tone,* New York, 1954, p. 20.

p. 128

As our age   Marshall McLuhan, *The Gutenberg Galaxy,* Toronto, 1962, p. 72.

p. 129

object for human   Pierre Schaeffer, "Music and Computers," *Music and Technology,*
Paris, 1970, p. 84.

p. 130

A composed   Ibid., p. 84.

The sound object   Pierre Schaeffer, *Trois Microsillons d'Examples Sonores,* Paris, 1967,
paras. 73.1 and 2.

p. 132

The Sonic Environment   *Environment and Behaviour,* Vol. 1, No. 1, June, 1969,
pp. 49–70.

CHAPTER NINE: *Classification*

p. 133

Disintegrating   Barry Truax, "Soundscape Studies: An Introduction to the World
Soundscape Project," *Numus West,* Vol. 5, 1974, p. 37.

p. 134

solfège des objets   *Traité des Objets Musicaux,* Paris, 1966, pp. 584–587.

CHAPTER TEN: *Perception*

p. 151

Arnheim and Gombrich   The works I am referring to are Arnheim's *Art and Visual
Perception* (Los Angeles, 1967) and Gombrich's *Art and Illusion* (New York,
1960).

p. 152

striking parallels   Cf. George A. Miller, *Language and Communication,* New York,
1951, pp. 70–71.

p. 153

sound-association tests   For instance, H. A. Wilmer's "An Auditory Sound Associa-
tion Technique," *Science,* 114, 1951, pp. 621–622; or D. R. Stone's "A
Recorded Auditory Apperception Test as a New Projective Technique," *The
Journal of Psychiatry,* 29, 1950, pp. 349–353.

sonological competence   Dr. Otto Laske, personal communication. See also reference
in Glossary of Soundscape Terms, p. 271.

p. 154

Time after time   Alan Edward Beeby, *Sound Effects on Tape,* London, 1966,
pp. 48–49.

p. 156

one of the subjects   Georg von Békésy, *Experiments in Hearing,* New York,
1960, p. 6.

Faced with   Alan Edward Beeby, *op. cit.,* p. 12.

p. 157

I know of Edmund Carpenter, *Eskimo,* Toronto, 1959, p. 27.

Auditory Ibid., p. 26.

p. 158

But other paths Iannis Xenakis, *Formalized Music,* Indiana, 1971, pp. 8–9.

p. 160

Soundmaking Peter F. Ostwald, *Soundmaking,* Springfield, 1963, pp. 119–124.

CHAPTER ELEVEN: *Morphology*

p. 161

Media Harold A. Innis, *Empire and Communications,* Oxford, 1950, p. 7.

p. 162

Then came the rigid Virgil, *Georgics,* Book I, lines 143–144, trans. C. Day Lewis, *The Eclogues and Georgics of Virgil,* New York, 1964. The original runs:

> *tun ferri rigor atque argutae lammina serrae*
> *(nam primi cuneis scindebant fissile lignum).*

stridor serrae Cicero, *Tusculan Disputations,* V, 40:116; and Lucretius, *On the Nature of Things,* II: 10.

p. 163

And the first *The Cantos of Ezra Pound,* London, 1954, p. 87.

Then came in sight Notker the Stammerer, *Life of Charlemagne,* trans. Lewis Thorpe, Harmondsworth, Middlesex, 1969, pp. 163–164.

p. 164

Mohammed warned Qur'an, Surah XXIV, vs. 31.

p. 165

The moon The example comes from J. F. Carrington's *Talking Drums of Africa,* New York, 1969, p. 33.

sixty miles According to E. A. Powell, in *The Map That Is Half Unrolled,* London, 1926, p. 128.

p. 166

In Mozart's day Professor Kurt Blaukopf, personal communication.

The siren Sir Henry Martin Smith, H. M. Chief Inspector of Fire Services; personal communication.

p. 167

two-tone horn Home Office, Fire Service Department, Specification No. JCDD/24, April, 1964.

Clang! *B. C. Saturday Sunset,* September 21, 1907, p. 13.

A long wolf howl "The Flame Fighters" by Garnett Weston in *British Columbia Magazine,* June, 1911, p. 562.

yelping siren Kenneth Laas, Federal Sign and Signal Corporation, Blue Island, Illinois; personal communication.

CHAPTER TWELVE: *Symbolism*

p. 169

A word or an image   Carl G. Jung, *Man and His Symbols*, New York, 1964, pp. 20–21.

Psychological Types   Carl G. Jung, *Psychological Types*, New York, 1924, p. 152.

p. 170

that state   W. H. Auden, *The Enchaféd Flood*, New York, 1967, pp. 6 and 13.

Water is   Carl G. Jung, *The Archetypes and the Collective Unconscious*, Princeton, N.J., 1968, pp. 18–19.

p. 171

he played the violin   Thomas Mann, *Stories of Three Decades*, New York, 1936, p. 87.

Man's descent   Carl G. Jung, *The Archetypes and the Collective Unconscious, op. cit.,* p. 17.

p. 172

Nature is   Novalis, *Schriften,* eds. P. Kluckhohn and R. Samuel, Stuttgart, Vol. 3, p. 452. I am grateful to Dr. Samuel for locating this quotation for me.

p. 174

The bell is hung   George M. Grant, *Ocean to Ocean, Sandford Fleming's Expedition Through Canada in 1872,* Toronto, 1873, p. 272.

The bell denotes   Durandus, Bishop of Mende, 1286. Quoted from *Tintinnabula,* Ernest Morris, London, 1959, pp. 43–44.

p. 175

The whole air   Emily Carr, *Hundreds and Thousands,* Toronto/Vancouver, 1966, pp. 248–249.

p. 177

Holy Rosary Cathedral   See *The Vancouver Soundscape,* Vancouver, 1974, and *Five Village Soundscapes,* Vancouver, 1976.

seven church bells   The account is to be found in Strindberg's *The Red Room,* New York/London, 1967, pp. 2–3.

center of the bell   I have borrowed this paragraph and numerous other ideas about the symbolism of the horn from an unpublished study of the subject by Bruce Davis.

p. 178

Greek word siren   Cf. Gabriel Germain, "The Sirens and the Temptation of Knowledge," in *Homer,* eds. G. Steiner and R. Fagles, New Jersey, 1962, p. 94.

p. 180

And, sitting in   Albert Camus, *The Outsider,* trans. Stuart Gilbert, Harmondsworth, Middlesex, 1972, pp. 98 and 104–105.

CHAPTER THIRTEEN: *Noise*

p. 184

some researchers   Alexander Cohen *et al.,* "Sociocusis—Hearing Loss from Non-Occupational Noise Exposure," *Sound and Vibration,* 4:11, November, 1970.

See also Clifford R. Bragdon, *Noise Pollution: The Unquiet Crisis,* Philadelphia, 1971, pp. 74–76.

power lawnmowers William A. Shearer, "Acoustical Threshold Shift from Power Lawnmower Noise," *Sound and Vibration,* 2:10, October, 1968.

rock concerts See *Time* magazine, August 9, 1968, p. 51.

Russian researchers "Séminaire Interrégional sur l'Habitat dans ses Rapports avec la Santé Public," World Health Organization PA/185.65. See summary in WHO *Chronicle,* October, 1966.

p. 185

Dr. Samuel Rosen Samuel Rosen *et al.,* "Presbycusis Study of a Relatively Noise-Free Population in the Sudan," American Otological Society, *Transactions,* Vol. 50, 1962.

traffic noise *A Community Noise Survey,* Greater Vancouver Regional District, 1971, p. 12.

p. 189

In those days *The Epic of Gilgamesh,* trans. N. K. Saunders, Harmondsworth, Middlesex, 1971, p. 105.

p. 190

By the thirteenth century A. L. Poole, ed., *Medieval England,* Vol. 1, Oxford, 1958, pp. 252–254.

p. 196

Venezuela Venezuela, *Gaceta Municipal,* Capitulo 1, articulo quinto (1972).

Tunis Ville de Tunis, *Arrêté du 17 Octobre, 1951,* art. 5.

p. 197

Voyage en Icarie Étienne Cabet, *Voyage en Icarie,* Paris, 1842, p. 65.

p. 198

In Luxembourg Ville de Luxembourg, *Règlement Général de Police,* Chapitre II, art. 32.

In Bonn Bonn, *Strassenordnung,* para. 5 (1970).

In Freiburg Freiburg, *Polizeiverordnung,* para. 2 (1968).

Between the hours Ville de Tunis, *Arrêté sur le Bruit,* article premier (1955).

one incident Albert Camus, *The Outsider,* trans. Stuart Gilbert, Harmondsworth, Middlesex, 1972, pp. 104–105.

schnitzels Dr. Hüblinger, Essen; personal communication.

mah-jong parties C. McGugan, Assistant to the Colonial Secretary, Hong Kong; personal communication.

hotels in India See S. K. Chatterjee, R. N. Sen and P. N. Saha, "Determination of the Level of Noise Originating from Room Air-Conditioners," *The Heating and Ventilating Engineer and Journal of Air-Conditioning,* Vol. 38, No. 59, February, 1965, pp. 429–433.

p. 199

Mombasa (Kenya) D. S. Obhrai, Town Clerk, Mombasa; personal communication.

Auckland (*New Zealand*) R. Agnew, Chief City Health Inspector, Auckland; personal communication.

Rabat (*Morocco*) Mohamed Sbith, Préfecture de Rabat-Salé; personal communication.

Izmir (*Turkey*) Izmir, *By-law Concerning Bus Terminals,* art. 25.

auctioneer's bells The repeal is contained in City of Melbourne, By-law 418 (1961).

No bell or crier Manila, Ordinance No. 1600, Sect. 846 (1961).

Manila *Ibid.*

Vendors *Ibid.,* Sect. 846-a.

Another ordinance Manila, Ordinance No. 4708, Sect. 848-a (1963).

Monday to Saturday Municipalidad de San Salvador, *Ley del Ramo Municipal,* art. 8 (1951).

The same sound Arthur Paulus, Ville de Luxembourg, Administration des travaux; personal communication.

p. 200

Manila restricts Manila, Ordinance No. 1600, Sect. 847 (1961).

In Chiclayo (*Peru*) Chiclayo, *Reglamento sobre Supresión de Ruidos Molestos en las Cuidades,* art. 11 (1957).

In Genoa Genoa, *Regolamenta di Polizia Comunale,* art. 64 (1969).

and Hartford Hartford, City Ordinance, 21–2k (1967).

Damascus (*Syria*) Damascus, By-law No. 401, Sect. 3, para. 8 (1950).

In Canada Cf. *A Survey of Community Noise By-laws in Canada (1972).* World Soundscape Project, Burnaby, B.C.

Adelaide (*Australia*) Adelaide, By-law No. IX, Sect. 3–1 (1937).

Buffalo (*New York*) Buffalo, Noise Control Ordinance, art. XVII, paras. 1703–11.

Sioux City (*Iowa*) Sioux City, Ordinance No. 21954, Sec. 9–11 (1972).

It shall be Oklahoma City, The Charter and General Code, Chapter 3, 9.3.09 (1960).

Salisbury (*Rhodesia*) Wm. Alves, Mayor of Salisbury; personal communication.

A person using I. A. McCutcheon, Town Clerk; personal communication.

p. 201

Noises associated Mary Douglas, "The Lele of Kasai," in Daryll Forde, ed., *African Worlds: Studies in the Cosmological Ideas and Social Values of African Peoples,* London, 1963, p. 12.

CHAPTER FOURTEEN: *Listening*

p. 207

Hare Krishna sect Cf. *Regina vs. Clay Harrold,* Vancouver Court of Appeal, March 19, 1971.

p. 212

David Lowenthal   David Lowenthal, "The American Scene," *The Geographical Review*, Vol. LVIII, No. 1, 1968, p. 72.

Mark Twain   Mark Twain, *Life on the Mississippi*, New York and London, 1929, pp. 79–80.

William James   William James, "On a Certain Blindness in Human Beings," in *Talks to Teachers on Psychology*, New York, 1958, pp. 149–169.

CHAPTER FIFTEEN: *The Acoustic Community*

p. 215

Between these ramparts   Notker the Stammerer, *Life of Charlemagne*, trans. Lewis Thorpe, Harmondsworth, Middlesex, 1969, p. 136.

p. 216

closed windows   Quoted from Kurt Blaukopf, *Hexenküche der Musik*, Teufen, Switzerland, 1959, p. 45.

p. 217

no echo answers   Quoted from David Lowenthal, "The American Scene," *The Geographical Review*, Vol. LVIII, No. 1, 1968, p. 71.

The dense forest   George Green, *History of Burnaby and Vicinity*, Vancouver, 1947, p. 22.

p. 219

One voice   Lucretius, *De Rerum Natura*, trans W. H. D. Rouse, London, 1924, pp. 289–291.

tenths of a second   W. C. Sabine, *Collected Papers on Acoustics*, New York, 1964, p. 170.

p. 221

It is probable   *Ibid.;* p. 255.

Hence in accordance   Vitruvius, *De Architectura*, Book V, trans. F. Granger, London, 1970, pp. 277–279.

Someone will say   *Ibid.*, p. 281.

p. 223

On the other hand   Leslie L. Doelle, *Environmental Acoustics*, New York, 1972, pp. 3, 6, 19–20.

CHAPTER SIXTEEN: *Rhythm and Tempo in the Soundscape*

p. 226

The heartbeat range   See the graph in Abraham Moles, *Information Theory and Esthetic Perception*, London, 1966, p. 139.

p. 228

in Proust   See Walter Benjamin, *Illuminations*, New York, 1969, p. 214.

He heard   Leo Tolstoy, *Anna Karenina*, trans. C. Garnett, New York, 1965, pp. 265 and 267.

p. 232
It is difficult   *Five Village Soundscapes,* Vancouver, 1976.

CHAPTER SEVENTEEN: *The Acoustic Designer*

p. 238
Gravity   Lao-tzu, *Tâo Teh King, The Texts of Taoism,* trans. James Legge, New York, 1962, p. 69.

Where things grow   Shên Tsung-ch'ien. Quoted from Jacques Maritain, *Creative Intuition in Art and Poetry,* Washington, 1953, p. 396.

p'ing and tsê   See John Hazedel Levis, *Foundations of Chinese Musical Art,* New York, 1964.

p. 239
Ballaarat, Australia   F. J. Rogers, Town Clerk, Ballaarat; personal communication.

p. 240
ice-making   Damascus, By-law No. 401, Sect. 3, para. 7 (1950).

p. 242
This simplification   P. Fortin, Ministère des Postes et Télécommunications, France; personal communication.

p. 244
We have also   Francis Bacon, *The New Atlantis,* London, 1906, pp. 294–295.

CHAPTER EIGHTEEN: *The Soniferous Garden*

p. 246
The gate   From "The Story of Nur-ed Din and Enis-El-Jelis," *The Thousand and One Nights,* New York, 1909, p. 222.

p. 247
From the Anio   Edith Wharton, *Italian Villas and Their Gardens,* New York, 1904, p. 144.

p. 248
In another garden   *The Diary of John Evelyn,* Vol. 1, ed. William Bray, London, 1901, p. 179.

Pneumatics   *The Pneumatics of Hero of Alexandria,* ed. Marie Boas Hall, London, 1971, pp. 31–32.

Vitruvius   Vitruvius, *De Architectura,* trans. F. Granger, London, 1970, Book X, p. 313.

hydraulic organs   *The Diary of John Evelyn, op. cit.,* p. 177.

p. 249
At the further   *Ibid.,* p. 52.

p. 250
I had had the weather   E. T. A. Hoffmann, *The Life and Opinions of Kater Murr,* trans. L. J. Kent and E. C. Knight, Chicago, 1969, p. 25.

This is a bow   A. C. Moule, "Musical and Other Sound-Producing Instruments of the Chinese," *Journal of the North-China Branch of the Royal Asiatic Society,* Vol. XXXIX, 1908, pp. 105–106.

p. 251

We heard   J. S. Brandtsbuys, "Music Among the Madurees," *Djava,* Vol. 8, 1928, p. 69.

CHAPTER NINETEEN: *Silence*

p. 253

Leaning on our   Nikos Kazantzakis, *Report to Greco,* New York, 1965, pp. 198–199.

p. 254

War Remembrance   Trans. Barry Truax, *Utrechts Stadsbad,* May 2, 1973, p. 3.

p. 256

Le silence   Blaise Pascal, *Pensées,* ed. Ch. M. des Granges, Paris, 1964, p. 131.

When I described   John Cage, *Silence,* Middletown, Connecticut, 1961, p. 8.

There is no such   *Ibid.,* p. 191.

p. 257

The analyst   Theodor Reik, *Listening with the Third Ear,* New York, 1948, pp. 122–123.

Whereof one   Ludwig Wittgenstein, *Traktatus,* London, 1922, remark 7.

p. 258

Give up haste   Lao-tzu, *Tâo Teh King, The Texts of Taoism,* Part II, Chapter 56, verse 2.

p. 259

Keep silence   Jalal-ud-din Rumi, *Divan i Shams i Tabriz.*

The essence   Kirpal Singh, *Naam or Word,* Delhi, 1970, p. 59.

EPILOGUE: *The Music Beyond*

p. 260

It forms   Alain Daniélou, *The Raga-s of Northern Indian Music,* London, 1968, p. 21.

How indeed   Boethius, *De Institutione Musica.* Quoted from *Source Readings in Music History,* Oliver Strunk, New York, 1950, p. 84.

INDEX

A Note About the Author

R. Murray Schafer was born in Sarnia, Ontario, in 1933. He attended the Royal Conservatory of Music in Toronto, studied in Austria and England, and served as Professor of Communication Studies at Simon Fraser University in British Columbia. As a composer, Mr. Schafer has written music in all forms, much of it experimental in nature. He has received several awards and grants for his musical work, including grants from the Canada Council, the Fromm Music Foundation, the Koussevitzky Music Foundation, as well as a Guggenheim fellowship. Mr. Schafer now lives near Bancroft, Ontario.

A Note About the Type

The text of this book was set in a computer version of Palatino, a type face designed by the noted German typographer Hermann Zapf. Named after Giovanbattista Palatino, a writing master of Renaissance Italy, Palatino was the first of Zapf's type faces to be introduced to America. The first designs for the face were made in 1948, and the fonts for the complete face were issued between 1950 and 1952. Like all Zapf-designed type faces, Palatino is beautifully balanced and exceedingly readable.

The book was composed by Datagraphics Press, Inc., Phoenix, Arizona; printed and bound by The Haddon Craftsmen, Inc., Scranton, Pennsylvania.

The book was designed by Earl Tidwell.